POLLUTION
CONTROL
PROBLEMS
AND
RELATED
FEDERAL
LEGISLATION

Edited by

Thomas J. Morrisey

State University College at Buffalo

MSS Information Corporation
655 Madison Avenue, New York, N.Y. 10021

This is a custom-made book of readings prepared for the courses taught by the editor, as well as for related courses and for college and university libraries. For information about our program, please write to:

MSS INFORMATION CORPORATION
655 Madison Avenue
New York, New York 10021

MSS wishes to express its appreciation to the authors of the articles in this collection for their cooperation in making their work available in this format.

Library of Congress Cataloging in Publication Data

Morrisey, Thomas J
    Pollution control problems and related Federal legislation.

    1. Environmental law — United States. I. Title.
KF3775.M67   344'.73'046   74-7152
ISBN 0-8422-5175-8
ISBN 0-8422-0418-0 (pbk.)

CONTENTS

INTRODUCTION

Pollution can be generally classified into four primary types: air, land (solid wastes), noise, and water pollution. Air pollution ranks number one in terms of the greatest concern by the public, industry, and the Federal government. Land pollution, frequently referred to as the third pollution, is becoming a serious problem as more and more solid wastes are generated in our economy. Noise pollution, one of the more recently identified types of pollution, has been recognized as a serious health hazard. Water pollution, long recognized as a problem, has been receiving more and more attention through new and strengthened legislation at the Federal and State levels. Usually, we are confronted with these four types of pollution daily.

This manual covers the problems and legislation related to air pollution control, solid waste management, noise, and clean water. In addition, the pollution control problems of pesticides and radiation are also covered. Inasmuch as different types of pollution control legislation have had a direct or indirect effect upon the economy in one form or another, a section outlining the energy problem is included for discussion and reference purposes. Pollution control legislation relating directly to the energy problem is covered.

T. J. Morrisey

toward a new environmental ethic

History may well record that the beginning of the decade of the seventies marked the turning point in man's use and abuse of the precious planet of which he is both ward and guardian. Governments at all levels geared up for an attack on environmental problems. Citizens from every walk of life mobilized themselves to defend the environment for themselves and their children. Industries, great and small, began to embrace a new view of corporate responsibility for protecting the environment. International bodies began planning in earnest for global control of global pollution.

In our own country, the National Environmental Policy Act, signed into law on January 1, 1970, established a national policy to "maintain conditions under which man and nature can exist in productive harmony, and fulfill the social, economic, and other requirements of present and future generations of Americans."

Under the new law, the President appointed a Council on Environmetal Quality to coordinate environmental matters at the federal level and serve as his principal advisors in such matters.

All federal actions and proposals which could have significant impact on the environment were made subject to review by federal, state, and local environmental authorities.

In 1970, by order of the President with the consent of the Congress, federal programs dealing

TOWARD A NEW ENVIRONMENTAL ETHIC, U.S. Government Printing Office: 1971-O — 443-062, pp. 2-9, 25.

7

with the environment were reorganized and strengthened.

The National Oceanic and Atmospheric Administration was established in the Department of Commerce and made responsible for research on long-range effects of pollution on the physical environment, especially global trends affecting the oceans and the atmosphere.

The United States Environmental Protection Agency, reporting directly to the President, was charged with mounting an integrated, coordinated attack on pollution, filling the need, in the President's words, for "a strong independent agency" to serve as an objective, impartial arbiter of environmental matters, particularly in establishing and enforcing pollution control standards.

Federal anti-pollution laws were strengthened and appropriations increased in 1970. Old pollution control laws were being discovered, and new laws were under development.

But, of course, public policy decisions affecting the environment are not all made in Washington. All over America change is taking place—change that challenges the pessimistic view that man is helpless to control the technological forces he has set in motion, change growing out of a new ecological perspective, translating a new environmental ethic into environmental action.

More and more states and municipalities are adopting environmental policies and programs which reflect a realistic understanding of ecologic relationships and needs.

The concept of regional bodies to deal with environmental matters that overlap existing jurisdictional boundaries is gaining acceptance and effectiveness.

Industry, more and more, is demonstrating a desire to overcome obstacles to pollution control and to make a positive contribution to environmental quality.

People all over the country are insisting that

we abandon the psychology of the blind bulldozer, that we refrain from paving over the whole world, that we stop polluting the air and the water and the earth. They are taking their cases to the courts. And their voices are being heard in corporate board rooms and by government officials and legislators at every level.

This last is the most important of all—the commitment of citizens of all ages and from all walks of life to environmental sanity. For the decisions that are being made today, and that will have to be made in the future, to preserve the environment, are truly decisions for our whole society.

The choice is ours. Americans are the privileged members of the free institutions of the richest country on earth. If we have the will, if we make sensible choices, we can build the kind of world we want—for ourselves and for those who will come after us.

The United States Environmental Protection Agency was established December 2, 1970, bringing together for the first time in a single agency the major environmental control programs of the Federal government. EPA is charged with mounting an integrated, coordinated attack on the environmental problems of air and water pollution, solid wastes management, pesticides, radiation, and noise.

To insure that the Agency is truly responsive to environmental needs in every part of the country, it has established a strong field organization, with Regional Offices located at ten major cities. The Regional Offices are staffed by specialists in each program area and headed by a Regional Administrator possessing broad authority to act for EPA in matters within his jurisdiction.

EPA's creation marked the end of the piece-meal approach to our nation's environmental problems which has, so often in the past, inhibited progress—or merely substituted one form of pollution for another.

EPA was created to lead a broad, comprehensive attack on pollution, and its administrative organization has been designed to make this approach a reality. The new organizational structure makes it easier to identify, and to take into account, all the factors bearing on pollution and its control. It makes it possible to resolve competing or conflicting program aims. It is facilitating the development of better information on the total impact of stresses reaching man from various parts of the environment, and makes possible more sensible choices about what constitutes a healthful, satisfying milieu for human life.

Most important of all, perhaps, EPA gives to the American people a single, independent, impartial agency to serve as their advocate for a livable environment.

standards-setting and enforcement

The United States Environmental Protection Agency is, first and foremost, a regulatory agency, with responsibilities for establishing and enforcing environmental standards, within the limits of its various statutory authorities.

Establishment of standards is central to the whole pollution control effort, for it is in this way that we define what each of us may and may not do to the environment on which we all depend.

Whatever we do, however careful we may be, we cannot avoid altering and being altered by the world about us. We are required, moreover, as was the first man, to use our human skills and ingenuity to convert the resources of the earth into the sustenance of human life.

The key, then, to sensible environmental control is to determine, within the limits of our knowledge, what changes are tolerable or useful and which must be banned or limited as harmful.

The standards set by EPA (in some cases, in cooperation with the States) have the force of law. They define the kinds of levels of pollutants which must be prevented from entering our air and water, and establish time-tables for achieving the prescribed quality. They set limits on radiation emissions and pesticide residues. Enforcement of environmental standards is, under certain laws, shared with the States, the Federal government acting only when the State fails to do so; in other instances, the Federal government has primary enforcement authority.

research and monitoring

EPA is also a research body, monitoring and analyzing the environment and conducting scientific studies into the causes and effects of pollution, the techniques of pollution control, and the environmental consequences of man's actions.

Effective action, particularly in standards-setting and enforcement, requires that EPA have sound data on what is being introduced into the environment, its impact on ecological stability, on human health, and on other factors important to human life. By close coordination of its various research programs, EPA strives to develop a synthesis of knowledge from the biological, physical and social sciences which can be interpreted in terms of total human and environmental needs.

Major aims of the Agency's research efforts at this time include:

- Expansion and improvement of environmental monitoring and surveillance to provide baselines of environmental quality.

- Advancement of understanding of long-term exposures to contaminants, of sub-acute or delayed effects on human and other organisms, of the combined and synergistic actions of chemical, biological, and physical stresses.

- Acceleration of progress in applied research into the control of pollutants, the recycling of so-called "wastes," and the development of sophisticated, non-polluting production processes.

- Improved assessment of trends of technical and social change and potential effects—first, second, and even third-order effects—on environmental quality.

11

• Improved understanding of the transport of materials through the environment; their passage through the media of air, water, and land; their ability to cross the various interfaces; and their various changes of state that can make them innocuous at one point and hazardous at another.

In addition to performing research in its own laboratories in various locations throughout the country, EPA, through grants and contracts, supports the studies of scientists in universities and other research institutions. The Agency also consolidates and evaluates information as it is developed throughout the scientific community to develop the best possible scientific *base* for environmental action.

technical and financial assistance

EPA serves also as a catalyst for environmental protection efforts at all levels of government by providing technical and financial assistance to state, regional, and local jurisdictions.

EPA publishes and gives wide distribution to its technical and scientific findings in all program areas, to advance the total body of scientific knowledge and hasten the application of new, proven pollution-control techniques.

Its "Technology Transfer" program is specifically designed to bridge the gap between the development and application of new techniques to control pollution. Workshops and seminars are held for state and local officials, design engineers, industrial representatives, and the public to introduce them to new, practicable control technology; technical bulletins and design manuals are widely disseminated.

Through EPA's ten Regional Offices, prompt assistance is given to State and local authorities, industries, and citizens in the solution of technical problems.

In several program areas, Federal funds are made available for the construction and operation of facilities to reduce pollution and to demonstrate new technology. Financial assistance is also provided for state and local governments to aid their environmental control programs.

manpower development

EPA provides training both in its own extensive training facilities and in universities and other educational institutions, to help develop the highly skilled manpower the Nation needs to combat environmental problems. Technical training in control techniques and program management is given to employees of state and local governments, industry, and other organizations. Support is given to universities for environmental courses. Fellowships are available to qualified students for advanced training.

citizens technology assessment

EPA also serves as a source of information to the public. By widely disseminating scientific data bearing on environmental problems, it tries to bring to concerned Americans the facts on which they, individually and in community, can make sound, rational choices in environmental issues.

The decisions that shape the world we live in are being made daily, by thousands of people, in homes and factories, in town councils and zoning boards, at the ballot box and in the shopping center.

Each of us, every day, is making his own "technology assessment." We may not always make the correct choice between competing benefits and costs, but we can, if we have the facts, make a sensible choice.

the environment — a new ingredient

EPA is not, by any means, an environmental czar. For one thing, it shares many of its enforcement authorities with the States, in accordance with principles and procedures established by the

13

Congress in the legislation governing its activities.

Moreover, since "the environment," after all, is practically everything, it follows that many agencies of government conduct activities that directly affect it. The Department of Transportation, for example, is concerned with highways, railroads, and air transport. The Department of Interior administers public lands and natural resources. The Departments of Housing and Urban Development; Agriculture; Health, Education, and Welfare; Defense—all carry out activities that are of the greatest importance in determining not only the kind of life we live but also the kind of world we live it in.

Under the National Environmental Policy Act, all departments of government are required to take into account the environmental impact of proposed actions and these are subjected to careful scrutiny before action can be undertaken.

to add it all up

Our Nation has resolved to "maintain conditions under which man and nature can exist in productive harmony." The United States Environmental Protection Agency has a key role to play in carrying out that National policy.

EPA is determined to be an advocate for the environment wherever it can, whenever it can, as decisions about our Nation's future are being made—whether it be in the councils of government, in the boardrooms of industry, or in the living rooms of our citizens.

review of federal activities

To ensure full consideration of environmental factors in Federal decision-making, each Federal agency is required to submit to the President's Council on Environmental Quality an *environmental impact statement* on any proposal for legislation or other major action significantly affecting the quality of the human environment. This must include:

- the environmental impact of the proposed action,
- any adverse environmental effects which cannot be avoided should the proposal be implemented,
- alternatives to the proposed action,
- the relationship between local short-term uses of man's environment and the maintenance and enhancement of long-term productivity, and
- any irreversible and irretrievable commitments of resources which would be involved in the proposed action should it be implemented.

Before filing with the Council, the statement must be circulated in draft to EPA and other appropriate federal, state, and local environmental agencies for their comments. These must accompany the final statement.

The final environmental impact statement, together with all comments, must be made available to the Congress and the public by the originating agency. Federal agencies must insure the fullest practicable provision of timely public information and understanding of the environmental impact of federal plans and programs including, whenever appropriate, public hearings.

EPA is specifically charged with making public its written comments on environmental impact statements and with publishing its determinations when these hold that a proposal is unsatisfactory from the standpoint of public health or welfare or environmental quality.

The Council on Environmental Quality considers all the evidence and advises the President as to the best course of action.

on-going programs

On-going programs transferred to EPA under the President's Reorganization Plan No. 3 included the functions of:

The Federal Water Quality Administration—from the Department of Interior

The National Air Pollution Control Administration

The Bureau of Solid Waste Management

The Bureau of Water Hygiene

The Bureau of Radiological Health (environmental radiation programs)

—from the Department of Health, Education, and Welfare

The Federal Radiation Council

Also transferred to EPA were responsibilities and authorities for:

Establishing standards for environmental chemicals—from the Atomic Energy Commission

Establishing tolerances for pesticide chemicals—from the Department of Health, Education, and Welfare

Registration and labeling of pesticides—from the Department of Agriculture

Conducting research on pesticides—from several Departments

Conducting research on ecological systems—from the Council on Environmental Quality

In addition, the Noise Abatement and Control Act of 1970 assigned to EPA the responsibility for studying the problem of noise and making recommendations for control.

research facilities

EPA's research is carried out through three National Environmental Research Centers located at Cincinnati, Ohio; Research Triangle Park, North Carolina; and Corvallis, Oregon. In addition, EPA maintains the Western Environmental Research Laboratory in Las Vegas, Nevada, for extensive research in radiological health and related safety programs.

The Centers direct and coordinate the work of satellite laboratories in various parts of the country.

The programs conducted at each Center cover a wide range of problems in all aspects of pollution and pollution control. However, the Cincinnati Center gives particular emphasis to pollution control methods; the North Carolina Center to the health effects of environmental factors; and the Corvallis Center to ecological effects.

Laboratory locations include:
Dauphin Island, Alabama
Montgomery, Alabama
College, Alaska
Gulf Breeze, Florida
Perrine, Florida
Athens, Georgia
Chamblee, Georgia
Rockville, Maryland
Ann Arbor, Michigan
Grosse Ile, Michigan
Duluth, Minnesota
Las Vegas, Nevada
Ada, Oklahoma
Narragansett, Rhode Island
Bears Bluff, South Carolina
Gig Harbor, Washington

air pollution control

OBJECTIVES:

The student should be able to identify the specific purposes of the Clean Air Act.

The student should be able to describe those economic, political, and social factors leading to the passage of the Clean Air Act.

The student should be able to describe the effect the Clean Air Act has had on various types of industries.

The student should be able to identify the strengths and weaknesses of the Clean Air Act as they relate to current pollution control problems and energy conservation measures.

INTRODUCTION:

In the United States air pollution is a problem in all large cities and in many small towns.

Each year over 200 million tons of manmade waste products are released into the air of the United States. About half of this pollution is produced as a result of the transportation system, coming chiefly from the internal-combustion engine.

In terms of weight—which is not necessarily in all cases the best indication of their importance—according to 1969 estimates 51 percent of these pollutants come from transportation sources, 16 percent from fuel combustion in stationary sources, 15 percent from industrial processes, 4 percent from solid waste disposal practices and 14 percent from forest fires and other miscellaneous sources. The main classes of primary pollutants include sulfur oxides, particulate matter, carbon monoxide, hydrocarbons and nitrogen oxides. Numerous other noxious gases and harmful particulates also are introduced into the at-

TOWARD A NEW ENVIRONMENTAL ETHIC, U.S. Government Printing Office: 1971-O — 443-062, pp. 11-12.

mosphere from a variety of specific activities. Photochemical oxidants, a category of secondary pollutants of extreme importance, are formed in the atmosphere when, under the influence of sunlight, nitrogen oxides combine with gaseous hydrocarbons.

At levels commonly found in urban areas, air pollution contributes to the incidence of such chronic ailments as emphysema, bronchitis, and asthma; diseases which have increased dramatically in recent decades.

Moreover, chemical and radiological substances produced by modern technology may threaten our health and the health of future generations in ways that we are far from fully understanding. Scientists are now beginning to give attention to such matters as the capacity of chemical agents in the atmosphere to produce mutagenic effects in biological systems, the metabolism of absorbed pollutants, the ways in which pollutants may alter the normal biochemistry of cells, affect the hormonal system, and alter the general functions of body activity.

The adverse economic effects of air pollution are much more varied and substantial than is generally realized. They range from the waste of fuel and other valuable resources, through the soiling and corrosion of physical structures of all kinds, to damage to agriculture and forests. Moreover, by reducing visibility, air pollution contributes to the toll of accidents in both air and ground travel.

Within the last decade we have begun to understand that air pollution is a complex phenomenon of global significance. It involves gaseous as well as particulate contaminants; both can sometimes be altered and rendered more hazardous through interreactions which occur in the atmosphere under the influence of sunlight, moisture and other environmental factors.

The first Federal program on air pollution was developed in 1955 when the Public Health Service conducted a modest air pollution research program and offered technical assistance to state and local governments, which traditionally have had primary responsibility for dealing with community air pollution problems.

In 1963 Congress passed the landmark Clean

Air Act. This law authorized financial assistance to state and local governments for the initiation and improvement of control programs, federal-interstate abatement actions, and the publication of criteria describing the effects of pollution. The law placed special emphasis on gaseous pollutants, particularly exhaust emissions from motor vehicles, and sulfur oxides from stationary sources.

In 1965, amendments to the Clean Air Act gave the federal program authority to curb motor vehicle emissions. Federal standards were first applied to 1968 model motor vehicles.

The Air Quality Act of 1967 called for a new and more comprehensive approach to the problem. It required the designation of air quality regions on the basis of meterologic and urban factors, and the publication of criteria documents (describing the effects of pollutants) accompanied by related documents on the types and costs of techniques available to carry out source control. Armed with these data, Governors were required to establish air quality standards and implementation plans for regions designated. The work accomplished under the 1967 legislation paved the way for enactment of the Clean Air Act Amendments of 1970 which were signed into law on December 31, 1970.

Under the provisions of the Clean Air Act Amendments of 1970:

• EPA established in 1971 *national ambient air quality standards* specifying the maximum levels to be permitted in the ambient air of the six principal and most widespread classes of air pollutants: particulate matter, sulfur oxides, hydrocarbons, carbon monoxide, photochemical oxidants, and nitrogen oxides. These comprise *primary* standards, required to protect the public health, and *secondary* standards (requiring further reductions in particulates and sulfur oxides), to prevent the many other undesirable effects of pollution.

• States must carry out approved *implementation plans* for limiting the emission of pollutants so as to achieve the *primary* standards by mid-1975 and to achieve the *secondary* standards within a reasonable period of time. If any State should fail to develop or carry out such plans, EPA is authorized to do so.

- EPA establishes and enforces *performance standards* (emission standards) limiting emissions from new or modified stationary sources of pollutants. The first such performance standards issued cover large steam-electric generating plants, municipal incinerators, cement factories, and sulfuric and nitric acid plants.

- EPA establishes and enforces Federal *emission standards* for pollutants that, while not necessarily widespread, are exceptionally hazardous to human health. Standards limiting emissions of beryllium, mercury, and asbestos are being promulgated in 1971.

- EPA establishes and enforces *emission standards* for new motor vehicles. Standards have been promulgated requiring a reduction of 90 percent in hydrocarbons and carbon monoxide emitted by 1975 models as compared with the 1970 requirements, and a 90 percent reduction in oxides of nitrogen by 1976.

- EPA may regulate or prohibit the manufacture or sale of fuels or fuel additives that result in harmful emissions or interfere with motor vehicle pollution control devices. The first such regulations will cover alkyl lead.

- Research is being stepped up and incentives are provided to encourage the early development of low-polluting motor vehicle propulsion systems, including government purchase and use of vehicles employing such systems.

- EPA is investigating the effects and control of aircraft emissions, and will publish, in September 1971, *emission standards* for aircraft, to be enforced by the Secretary of Transportation.

- Citizens are specifically authorized to take civil court action against private or governmental officials failing to carry out the provisions of the law. Public hearings are required at various steps in the standards-setting, enforcement, and regulatory procedures to enable all interested persons to make their feelings known.

- EPA conducts extensive *research* into all aspects of air pollution, both in its own laboratories and through grants and contracts. It constructs and operates *demonstration plants* or processes, or financially assists such projects.

- *Financial grants* are made to state, interstate, and local agencies to aid their air pollution control programs.

20

THE CLEAN AIR ACT

"TITLE I—AIR POLLUTION PREVENTION AND CONTROL [1]

"FINDINGS AND PURPOSES

"SEC. 101. (a) The Congress finds—

"(1) that the predominant part of the Nation's population is located in its rapidly expanding metropolitan and other urban areas, which generally cross the boundary lines of local jurisdictions and often extend into two or more States;

"(2) that the growth in the amount and complexity of air pollution brought about by urbanization, industrial development, and the increasing use of motor vehicles, has resulted in mounting dangers to the public health and welfare, including injury to agricultural crops and livestock, damage to and the deterioration of property, and hazards to air and ground transportation;

"(3) that the prevention and control of air pollution at its source is the primary responsibility of States and local governments; and

"(4) that Federal financial assistance and leadership is essential for the development of cooperative Federal, State, regional, and local programs to prevent and control air pollution.

"(b) The purposes of this title are—

"(1) to protect and enhance the quality of the Nation's air resources so as to promote the public health and welfare and the productive capacity of its population;

"(2) to initiate and accelerate a national research and development program to achieve the prevention and control of air pollution;

"(3) to provide technical and financial assistance to State and local governments in connection with the development and execution of their air pollution prevention and control programs; and

"(4) to encourage and assist the development and operation of regional air pollution control programs.

"COOPERATIVE ACTIVITIES AND UNIFORM LAWS

"SEC. 102. (a) The Administrator shall encourage cooperative activities by the States and local governments for the prevention

[1] Clean Air Act (42 U.S.C. 1857 et seq.) includes the Clean Air Act of 1963 (P.L. 88–206), and amendments made by the "Motor Vehicle Air Pollution Control Act"—P.L. 89–272 (October 20, 1965), the "Clean Air Act Amendments of 1966—P.L. 89–675 (October 15, 1966), the "Air Quality Act of 1967"—P.L. 90–148 (November 21, 1967), and the "Clean Air Amendments of 1970"—P.L. 91–604—(December 31, 1970).

THE CLEAN AIR ACT, December 1970, U.S. Government Printing Office: 1971 O — 413-241, pp. 1-56.

and control of air pollution; encourage the enactment of improved and, so far as practicable in the light of varying conditions and needs, uniform State and local laws relating to the prevention and control of air pollution; and encourage the making of agreements and compacts between States for the prevention and control of air pollution.

"(b) The Administrator shall cooperate with and encourage cooperative activities by all Federal departments and agencies having functions relating to the prevention and control of air pollution, so as to assure the utilization in the Federal air pollution control program of all appropriate and available facilities and resources within the Federal Government.

"(c) The consent of the Congress is hereby given to two or more States to negotiate and enter into agreements or compacts, not in conflict with any law or treaty of the United States, for (1) cooperative effort and mutual assistance for the prevention and control of air pollution and the enforcement of their respective laws relating thereto, and (2) the establishment of such agencies, joint or otherwise, as they may deem desirable for making effective such agreements or compacts. No such agreement or compact shall be binding or obligatory upon any State a party thereto unless and until it has been approved by Congress. It is the intent of Congress that no agreement or compact entered into between States after the date of enactment of the Air Quality Act of 1967, which relates to the control and abatement of air pollution in an air quality control region, shall provide for participation by a State which is not included (in whole or in part) in such air quality control region.

"RESEARCH, INVESTIGATION, TRAINING, AND OTHER ACTIVITIES

"SEC. 103. (a) The Administrator shall establish a national research and development program for the prevention and control of air pollution and as part of such program shall—

"(1) conduct, and promote the coordination and acceleration of, research, investigations, experiments, training, demonstrations, surveys, and studies relating to the causes, effects, extent, prevention, and control of air pollution;

"(2) encourage, cooperate with, and render technical services and provide financial assistance to air pollution control agencies and other appropriate public or private agencies, institutions, and organizations, and individuals in the conduct of such activities;

"(3) conduct investigations and research and make surveys concerning any specific problem of air pollution in cooperation with any air pollution control agency with a view to recommending a solution of such problem, if he is requested to do so by such agency or if, in his judgment, such problem may affect any community or communities in a State other than that in which the source of the matter causing or contributing to the pollution is located;

"(4) establish technical advisory committees composed of

recognized experts in various aspects of air pollution to assist in the examination and evaluation of research progress and proposals and to avoid duplication of research.

"(b) In carrying out the provisions of the preceding subsection the Administrator is authorized to—

"(1) collect and make available, through publications and other appropriate means, the results of and other information, including appropriate recommendations by him in connection therewith, pertaining to such research and other activities;

"(2) cooperate with other Federal departments and agencies, with air pollution control agencies, with other public and private agencies, institutions, and organizations, and with any industries involved, in the preparation and conduct of such research and other activities;

"(3) make grants to air pollution control agencies, to other public or nonprofit private agencies, institutions, and organizations, and to individuals, for purposes stated in subsection (a) (1) of this section;

"(4) contract with public or private agencies, institutions, and organizations, and with individuals, without regard to sections 3648 and 3709 of the Revised Statutes (31 U.S.C. 529; 41 U.S.C. 5);

"(5) provide training for, and make training grants to, personnel of air pollution control agencies and other persons with suitable qualifications;

"(6) establish and maintain research fellowships, in the Environmental Protection Agency and at public or nonprofit private educational institutions or research organizations;

"(7) collect and disseminate, in cooperation with other Federal departments and agencies, and with other public or private agencies, institutions, and organizations having related responsibilities, basic data on chemical, physical, and biological effects of varying air quality and other information pertaining to air pollution and the prevention and control thereof; and

"(8) develop effective and practical processes, methods, and prototype devices for the prevention or control of air pollution.

"(c) In carrying out the provisions of subsection (a) of this section the Administrator shall conduct research on, and survey the results of other scientific studies on, the harmful effects on the health or welfare of persons by the various known air pollutants.

"(d) The Administrator is authorized to construct such facilities and staff and equip them as he determines to be necessary to carry out his functions under this Act.

"(e) If, in the judgment of the Administrator, an air pollution problem of substantial significance may result from discharge or discharges into the atmosphere, he may call a conference concerning this potential air pollution problem to be held in or near one or more of the places where such discharge or discharges are occurring or will occur. All interested persons shall be given an opportunity to be heard at such conference, either orally or in writing, and shall be permitted to appear in person or by representative in accordance with porcedures prescribed by the Administrator. If ... the Administrator finds, on the basis of evidence presented at such conference, that the discharge or discharges if permitted to take place or continue are likely to cause or contribute to air pollution subject to abatement under section 115, he shall send such findings, together with recommendations concerning the measures which he finds reasonable and suitable to prevent such pollution, to the person or persons whose actions will result in the discharge or discharges involved; to air pollution agencies of the State or States and of the municipality or municipalities where such discharge or discharges will originate; and to the interstate air pollution control agency, if any, in the jurisdictional area of which any such municipality is located. Such findings and recommendations shall be advisory only, but shall be admitted together with the record of the conference, as part of the proceedings under subsections (b), (c), (d), (e), and (f) of section 115.

"(f) (1) In carrying out research pursuant to this Act, the Administrator shall give special emphasis to research on the short- and long-term effects of air pollutants on public health and welfare. In the furtherance of such research, he shall conduct an accelerated research program—

"(A) to improve knowledge of the contribution of air pollutants to the occurrence of adverse effects on health, including, but not limited to, behavioral, physiological, toxicological, and biochemical effects; and

"(B) to improve knowledge of the short- and long-term effects of air pollutants on welfare.

"(2) In carrying out the provisions of this subsection the Administrator may—

"(A) conduct epidemiological studies of the effects of air pollutants on mortality and morbidity;

"(B) conduct clinical and laboratory studies on the immunologic, biochemical, physiological, and the toxicological effects including carcinogenic, teratogenic, and mutagenic effects of air pollutants;

"(C) utilize, on a reimbursable basis, the facilities of existing Federal scientific laboratories and research centers;

"(D) utilize the authority contained in paragraphs (1) through (4) of subsection (b); and

"(E) consult with other appropriate Federal agencies to assure that research or studies conducted pursuant to this subsection will be coordinated with research and studies of such other Federal agencies.

"(3) In entering into contracts under this subsection, the Administrator is authorized to contract for a term not to exceed 10 years in duration. For the purposes of this paragraph, there are authorized to be appropriated $15,000,000. Such amounts as are appropriated shall remain available until expended and shall be in addition to any other appropriations under this Act."

"SEC. 104. (a) The Administrator shall give special emphasis to research and development into new and improved methods, having industrywide application, for the prevention and control of air pollution resulting from the combustion of fuels. In furtherance of such research and development he shall—

"(1) conduct and accelerate research programs directed toward development of improved, low-cost techniques for—

"(A) control of combustion byproducts of fuels,

"(B) removal of potential air pollutants from fuels prior to combustion,

"(D) control of emissions from the evaporation of fuels,

"(D) improving the efficiency of fuels combustion so as to decrease atmospheric emissions, and

"(E) producing synthetic or new fuels which, when used, result in decreased atmospheric emissions."

"(2) provide for Federal grants to public or nonprofit agencies, institutions, and organiations and to individuals, and contracts with public or private agencies, institutions, or persons, for payment of (A) part of the cost of acquiring, constructing, or otherwise securing for research and development purposes, new or improved devices or methods having industrywide application of improved devices or methods having industrywide application of preventing or controlling discharges into the air of various types of pollutants; (B) part of the cost of programs to develop low emission alternatives to the present internal combustion engine; (C) the cost to purchase vehicles and vehicle engines, or portions thereof, for research, development, and testing purposes; and (D) carrying out the other provisions of this section, without regard to sections 3648 and 3709 of the Revised Statutes (31 U.S.C. 529; 41 U.S.C. 5) : *Provided,* That research or demonstration contracts awarded pursuant to this subsection or demonstration contracts awarded pursuant to this subsection (including contracts for construction) may be made in accordance with, and subject to the limitations provided with respect to research contracts of the military departments in, section 2353 of title 10, United States Code, except that the determination, approval, and certification required thereby shall be made by the Administrator: *Provided further,* That no grant may be made under this paragraph in excess of $1,500,000;

"(3) determine, by laboratory and pilot plant testing, the results of air pollution research and studies in order to develop new or improved processes and plant designs to the point where they can be demonstrated on a large and practical scale;

"(4) construct, operate, and maintain, or assist in meeting the cost of the construction, operation, and maintenance of new or improved demonstration plants or processes which have promise of accomplishing the purposes of this Act;

"(5) study new or improved methods for the recovery and marketing of commercially valuable byproducts resulting from the removal of pollutants.

"(b) In carrying out the provisions of this section, the Administrator may—

"(1) conduct and accelerate research and development of low-cost instrumentation techniques to facilitate determination of quantity and quality of air pollutant emissions, including, but not limited to, automotive emissions;

"(2) utilize, on a reimbursable basis, the facilities of existing Federal scientific laboratories;

"(3) establish and operate necessary facilities and test sites at which to carry on the research, testing, development, and programing necessary to effectuate the purposes of this section;

"(4) acquire secret processes, technical data, inventions, patent applications, patents, licenses, and an interest in lands, plants, and facilities, and other property or rights by purchase, license, lease, or donation; and

"(5) cause on-site inspections to be made of promising domestic and foreign projects, and cooperate and participate in their development in instances in which the purposes of the Act will be served thereby.

"(c) For the purposes of this section there are authorized to be appropriated $75,000,000 for the fiscal year ending June 30, 1971, $125,000,000 for the fiscal year ending June 30, 1972, and $150,-000,000 for the fiscal year ending June 30, 1973. Amounts appropriated pursuant to this subsection shall remain available until expended.

"GRANTS FOR SUPPORT OF AIR POLLUTION PLANNING AND
CONTROL PROGRAMS

"SEC. 105. (a) (1) (A) The Administrator may make grants to air pollution control agencies in an amount up to two-thirds of the cost of planning, developing, establishing, or improving, and up to one-half of the cost of maintaining, programs for the prevention and control of air pollution or implementation of national primary and secondary ambient air quality standards.

"(B) Subject to subparagraph (C), the Administrator may make grants to air pollution control agencies within the meaning of paragraph (1), (2), or (4) of section 302(b) in an amount up to three-fourths of the cost of planning, developing, establishing, or improving, and up to three-fifths of the cost of maintaining, any program for the prevention and control of air pollution or implementation of national primary and secondary ambient air quality standards in an area that includes two or more municipalities, whether in the same or different States.

"(C) With respect to any air quality control region or portion thereof for which there is an applicable implementation plan under section 110, grants under subparagraph (B) may be made only to air pollution control agencies which have substantial responsibilities for carrying out such applicable implementation plan."

"(2) Before approving any grant under this subsection to any air pollution control agency within the meaning of sections 302 (b) (2) and 302(b) (4) the Administrator shall receive assurances that such agency provides for adequate representation of appropriate State, interstate, local, and (when appropriate) international, interests in the air quality control region.

"(3) Before approving any planning grant under this subsection to any air pollutant control agency within the meaning of sections 302(b) (2) and 302(b) (4), the Administrator shall receive assurances that such agency has the capability of developing a comprehensive air quality plan for the air quality control region, which plan shall include (when appropriate) a recommended system of alerts to avert and reduce the risk of situations in which there may be imminent and serious danger to the public health or welfare from air pollutants and the various aspects relevant to the establishment of air quality standards for such air quality control region, including the concentration of industries, other commercial establishments, population and naturally occurring factors which shall affect such standards.

"(b) from the sums available for the purposes of subsection (a) of this section for any fiscal year, the Administrator shall from time to time make grants to air pollution control agencies upon such terms and conditions as the Administrator may find necessary to carry out the purpose of this section. In establishing regulations for the granting of such funds the Administrator shall, so far as practicable, give due consideration to (1) the population, (2) the extent of the actual or potential air pollution problem, and (3) the financial need of the respective agencies. No agency shall receive any grant under this section during any fiscal year when its expenditures of non-Federal funds for other than nonrecurrent expenditures for air pollution control programs will be less than its expenditures were for such programs during the preceding fiscal year; and no agency shall receive any grant under this section with respect to the maintenance of a program for the prevention and control of air pollution unless the Administrator is satisfied that such grant will be so used as to supplement and, to the extent practicable, increase the level of State, local, or other non-Federal funds that would in the absence of such grant be made available for the maintenance of such program, and will in no event supplant such State, local, or other non-Federal funds. No grant shall be made under this section until the Administrator has consulted with the appropriate official as designated by the Governor or Governors of the State or States affected.

"(c) Not more than 10 per centum of the total of funds appropriated or allocated for the purposes of subsection (a) of this section shall be granted for air pollution control programs in any one State. In the case of a grant for a program in an area crossing State boundaries, the Administrator shall determine the portion of such grant that is chargeable to the percentage limitation under this subsection for each State into which such area extends.

"(d) The Administrator, with the concurrence of any recipient of a grant under this section, may reduce the payments to such recipient by the amount of the pay, allowances, traveling expenses, and any other costs in connection with the detail of any officer or employee to the recipient under section 301 of this Act, when such detail is for the convenience of, and at the request of, such recipient and for the purpose of carrying out the provisions of this Act. The amount by which such payments have been reduced shall be available for payment of such costs by the Administrator, but shall, for the purpose of determining the amount of any grant to a recipient under subsection (a) of this section, be deemed to have been paid to such agency.

"INTERSTATE AIR QUALITY AGENCIES OR COMMISSIONS

"SEC. 106. For the purpose of developing implementation plans for any interstate air quality control region designated pursuant to section 107, the Administrator is authorized to pay, for two years, up to 100 per centum of the air quality planning program costs of any agency designated by the Governors of the affected States, which agency shall be capable of recommending to the Governors plans for implementation of national primary and secondary ambient air quality standards and shall include representation from the States and appropriate political subdivisions within the air quality control region. After the initial two-year period the Administrator is authorized to make grants to such agency in an amount up to three-fourths of the air quality planning program costs of such agency.

"AIR QUALITY CONTROL REGIONS

"SEC. 107. (a) Each State shall have the primary responsibility for assuring air quality within the entire geographic area comprising such State by submitting an implementation plan for such State which will specify the manner in which national primary and secondary ambient air quality standards will be achieved and maintained within each air quality control region in such State.

"(b) For purposes of developing and carrying out implementation plans under section 110—

"(1) an air quality control region designated under this section before the date of enactment of the Clean Air Amendments of 1970, or a region designated after such date under subsection (c), shall be an air quality control region; and

"(2) the portion of such State which is not part of any such designated region shall be an air quality control region, but such portion may be subdivided by the State into two or more air quality control regions with the approval of the Administrator.

"(c) The Administrator shall, within 90 days after the date of enactment of the Clean Air Amendments of 1970, after consultation with appropriate State and local authorities, designate as an air quality control region any interstate area or major intrastate area which he deems necessary or appropriate for the attainment and maintenance of ambient air quality standards. The Administrator shall immediately notify the governors of the affected States of any designation made under this subsection.

"AIR QUALITY CRITERIA AND CONTROL TECHNIQUES

"SEC. 108. (a) (1) For the purpose of establishing national primary and secondary ambient air quality standards, the Administrator shall within 30 days after the rate of enactment of the Clean Air Amendments of 1970 publish, and shall from time to time thereafter revise, a list which includes each air pollutant—

"(A) which in his judgment has an adverse effect on public health and welfare;

"(B) the presence of which in the ambient air results from numerous or diverse mobile or stationary sources; and

"(C) for which air quality criteria had not been issued before the date of enactment of the Clean Air Amendments of 1970, but for which he plans to issue air quality criteria under this section.

"(2) The Administrator shall issue air quality criteria for an air pollutant within 12 months after he has included such pollutant in a list under paragraph (1). Air quality criteria for an air pollutant shall accurately reflect the latest scientific knowledge useful in indicating the kind and extent of all identifiable effects on public health or welfare which may be expected from the presence of such pollutant in the ambient air, in varying quantities. The criteria for an air pollutant, to the extent practicable, shall include information on—

"(A) those variable factors (including atmospheric conditions) which of themselves or in combination with other factors may alter the effects on public health or welfare of such air pollutant;

"(B) the types of air pollutants which, when present in the atmosphere, may interact with such pollutant to produce an adverse effect on public health or welfare; and

"(C) any known or anticipated adverse effects on welfare.

"(b) (1) Simultaneously with the issuance of criteria under subsection (a), the Administrator shall, after consultation with appropriate advisory committees and Federal departments and agencies, issue to the States and appropriate air pollution control agencies information on air pollution control techniques, which information shall include data relating to the technology and costs of emission control. Such information shall include such data as are available on available technology and alternative methods of prevention and control of air pollution. Such information shall also include data on alternative fuels, processes, and operating methods which will result in elimination of significant reduction of emissions.

"(2) In order to assist in the development of information on pollution control techniques, the Administrator may establish a standing consulting committee for each air pollutant included in a list published pursuant to subsection (a)(1), which shall be comprised of technically qualified individuals representative of State and local governments, industry, and the academic community. Each such committee shall submit as appropriate, to the Administrator information related to that required by in paragraph (1).

"(c) The Administrator shall from time to time review, and, as appropriate, modify, and reissue any criteria or information on control techniques issued pursuant to this section.

"(d) The issuance of air quality criteria and information on air pollution control techniques shall be announced in the Federal Register and copies shall be made available to the general public.

"NATIONAL AMBIENT AIR QUALITY STANDARDS

"SEC. 109. (a)(1) The Administrator—
   "(A) within 30 days after the date of enactment of the Clean Air Amendments of 1970, shall publish proposed regulations prescribing a national primary ambient air quality standard and a national secondary ambient air quality standard for each air pollutant for which air quality criteria have been issued prior to such date of enactment; and
   "(B) after a reasonable time for interested persons to submit written comments thereon (but no later than 90 days after the initial publication of such proposed standards) shall by regulation promulgate such proposed national primary and secondary ambient air quality standards with such modifications as he deems appropriate.

"(2) With respect to any air pollutant for which air quality criteria are issued after the date of enactment of the Clean Air Amendments of 1970, the Administrator shall publish, simultaneously with the issuance of such criteria and information, proposed national primary and secondary ambient air quality standards for any such pollutant. The procedure provided for in paragraph (1)(B) of this subsection shall apply to the promulgation of such standards.

"(b)(1) National primary ambient air quality standards, prescribed, under subsection (a) shall be ambient air quality standards the attainment and maintenance of which in the judgment of the Administrator, based on such criteria and allowing an adequate margin of safety, are requisite to protect the public health. Such primary standards may be revised in the same manner as promulgated.

"(2) Any national secondary ambient air quality standard prescribed, under subsection (a) shall specify a level of air quality the attainment and maintenance of which in the judgment of the

Administrator, based on such criteria, is requisite to protect the public welfare from any known or anticipated adverse effects associated with the presence of such air pollutant in the ambient air. Such secondary standards may be revised in the same manner as promulgated.

"IMPLEMENTATION PLANS

"SEC. 110. (a) (1) Each State shall, after reasonable notice and public hearings, adopt and submit to the Administrator, within nine months after the promulgation of a national primary ambient air quality standard (or any revision thereof) under section 109 for any air pollutant, a plan which provides for implementation, maintenance, and enforcement of such primary standard in each air quality control region (or portion thereof) within such State. In addition, such State shall adopt and submit to the Administrator (either as a part of a plan submitted under the preceding sentence or separately) within nine months after the promulgation of a national ambient air quality secondary standard (or revision thereof), a plan which provides for implementation, maintenance, and enforcement of such secondary standard in each air quality control region (or portion thereof) within such State. Unless a separate public hearing is provided, each State shall consider its plan implementing such secondary standard at the hearing required by the first sentence of this paragraph.

"(2) The Administrator shall, within four months after the date required for submission of a plan under paragraph (1), approve or disapprove such plan for each portion thereof. The Administrator shall approve such plan, or any portion thereof, if he determines that it was adopted after reasonable notice and hearing and that—

"(A) (i) in the case of a plan implementing a national primary ambient air quality standard, it provides for the attainment of such primary standard as expeditiously as practicable but (subject to subsection (e)) in no case later than three years from the date of approval of such plan (or any revision thereof to take account of a revised primary standard) ; and (ii) in the case of a plan implementing a national secondary ambient air quality standard, it specifies a reasonable time at which such secondary standard will be attained;

"(B) it includes emission limitations, schedules, and timetables for compliance with such limitations, and such other measures as may be necessary to insure attainment and maintenance of such primary or secondary standard, including, but not limited to, land-use and transportation controls;

"(C) it includes provision for establishment and operation of appropriate devices, methods, systems, and procedures necessary to (i) monitor, compile, and analyze data on ambient air quality and, (ii) upon request, make such data available to the Administrator;

"(D) it includes a procedure, meeting the requirements of paragraph (4), for review (prior to construction or modifica-

31

tion) of the location of new sources to which a standard of performance will apply;

"(E) it contains adequate provisions for intergovernmental cooperation, including measures necessary to insure that emissions of air pollutants from sources located in any air quality control region will not interfere with the attainment or maintenance of such primary or secondary standard in any portion of such region outside of such State or in any other air quality control region;

"(F) it provides (i) necessary assurances that the State will have adequate personnel, funding, and authority to carry out such implementation plan, (ii) requirements for installation of equipment by owners or operators of stationary sources to monitor emissions from such sources, (iii) for periodic reports on the nature and amounts of such emissions; (iv) that such reports shall be correlated by the State agency with any emission limitations or standards established pursuant to this Act, which reports shall be available at reasonable times for public inspection; and (v) for authority comparable to that in section 303, and adequate contingency plans to implement such authority;

"(G) it provides, to the extent necessary and practicable, for periodic inspection and testing of motor vehicles to enforce compliance with applicable emission standards; and

"(H) it provides for revision, after public hearings, of such plan (i) from time to time as may be necessary to take account of revisions of such national primary or secondary ambient air quality standard or the availability of improved or more expeditious methods of achieving such primary or secondary standard; or (ii) whenever the Administrator finds on the basis of information available to him that the plan is substantially inadequate to achieve the national ambient air quality primary or secondary standard which it implements.

"(3) The Administrator shall approve any revision of an implementation plan applicable to an air quality control region if he determines that it meets the requirements of paragraph (2) and has been adopted by the State after reasonable notice and public hearings.

"(4) The procedure referred to in paragraph (2)(D) for review, prior to construction or modification, of the location of new sources shall (A) provide for adequate authority to prevent the construction or modification of any new source to which a standard of performance under section 111 will apply at any location which the State determines will prevent the attainment or maintenance within any air quality control region (or portion thereof) within such State of a national ambient air quality primary or secondary standard, and (B) require that prior to commencing construction or modification of any such source, the owner or operator thereof shall submit to such State such information as may be necessary to permit the State to make a determination under clause (A).

"(b) The Administrator may, wherever he determines necessary, extend the period for submission of any plan or portion thereof which implements a national secondary ambient air quality standard for a period not to exceed eighteen months from the date otherwise required for submission of such plan.

"(c) The Administrator shall, after consideration of any State hearing record, promptly prepare and publish proposed regulations setting forth an implementation plan, or portion thereof, for a State if—

"(1) The State fails to submit an implementation plan for any national ambient air quality primary or secondary standard within the time prescribed,

"(2) the plan, or any portion thereof, submitted for such State is determined by the Administrator not to be in accordance with the requirements of this section, or

"(3) the State fails, within 60 days after notification by the Administrator or such longer period as he may prescribe, to revise an implementation plan as required pursuant to a provision of its plan referred to in subsection (a) (2) (H).

If such State held no public hearing associated with respect to such plan (or revision thereof), the Administrator shall provide opportunity for such hearing within such State on any proposed regulation. The Administrator shall, within six months after the date required for submission of such plan (or revision thereof), promulgate any such regulations unless, prior to such promulgation, such State has adopted and submitted a plan (or revision) which the Administrator determines to be in accordance with the requirements of this section.

"(d) For purposes of this Act, an applicable implementation plan is the implementation plan, or most recent revision thereof, which has been approved under subsection (a) or promulgated under subsection (c) and which implements a national primary or secondary ambient air quality standard in a State.

"(e) (1) Upon application of a Governor of a State at the time of submission of any plan implementing a national ambient air quality primary standard, the Administrator may (subject to paragraph (2)) extend the three-year period referred to in subsection (a) (2) (A) (i) for not more than two years for an air quality control region if after review of such plan the Administrator determines that—

"(A) one or more emission sources (or classes of moving sources) are unable to comply with the requirements of such plan which implement such primary standard because the necessary technology or other alternatives are not available or will not be available soon enough to permit compliance within such three-year period, and

"(B) the State has considered and applied as a part of its plan reasonably available alternative means of attaining such primary standard and has justifiably concluded that attainment of such primary standard within the three years cannot be achieved.

"(2) The Administrator may grant an extension under paragraph (1) only if he determines that the State plan provides for—

"(A) application of the requirements of the plan which implement such primary standard to all emission sources in such region other than the sources (or classes) described in paragraph (1) (A) within the three-year period, and

"(B) such interim measures of control of the sources (or classes) described in paragraph (1) (A) as the Administrator determines to be reasonable under the circumstances.

"(f) (1) Prior to the date on which any stationary source or class of moving sources is required to comply with any requirement of an applicable implementation plan the Governor of the State to which such plan applies may apply to the Administrator to postpone the applicability of such requirement to such source (or class) for not more than one year. If the Administrator determines that—

"(A) good faith efforts have been made to comply with such requirement before such date,

"(B) such source (or class) is unable to comply with such requirement because the necessary technology or other alternative methods of control are not available or have not been available for a sufficient period of time,

"(C) any available alternative operating procedures and interim control measures have reduced or will reduce the impact of such source on public health, and

"(D) the continued operation of such source is essential to national security or to the public health or welfare, then the Administrator shall grant a postponement of such requirement.

"(2) (A) Any determination under paragraph (1) shall (i) be made on the record after notice to interested persons and opportunity for hearing, (ii) be based upon a fair evaluation of the entire record at such hearing, and (iii) include a statement setting forth in detail the findings and conclusions upon which the determination is based.

"(B) Any determination made pursuant to this paragraph shall be subject to judicial review by the United States court of appeals for the circuit which includes such State upon the filing in such court within 30 days from the date of such decision of a petition by any interested person praying that the decision be modified or set aside in whole or in part. A copy of the petition shall forthwith be sent by registered or certified mail to the Administrator and thereupon the Administrator shall certify and file in such court the record upon which the final decision complained of was issued, as provided in section 2112 of title 28, United States Code. Upon the filing of such petition the court shall have jurisdiction to affirm or set aside the determination complained of in whole or in part. The findings of the Administrator with respect to questions of fact (including each determination made under subparagraphs (A), (B), (C), and (D) of paragraph (1)) shall be sustained if based upon a fair evaluation of the entire record at such hearing.

"(C) Proceedings before the court under this paragraph shall take precedence over all the other causes of action on the docket

and shall be assigned for hearing and decision at the earliest practicable date and expedited in every way.

"(D) Section 307 (a) (relating to subpenas) shall be applicable
to any proceeding under this subsection.

"SEC. 111. (a) For purposes of this section:

"(1) The term 'standard of performance' means a standard
for emissions of air pollutants which reflects the degree of
emission limitation achievable through the application of the
best system of emission reduction which (taking into account
the cost of achieving such reduction) the Administrator determines has been adequately demonstrated.

"(2) The term 'new source' means any stationary source,
the construction or modification of which is commenced after
the publication of regulations (or, if earlier, proposed regulations) prescribing a standard of performance under this
section which will be applicable to such source.

"(3) The term 'stationary source' means any building,
structure, facility, or installation which emits or may emit
any air pollutant.

"(4) The term 'modification' means any physical change in,
or change in the method of operation of, a stationary source
which increases the amount of any air pollutant emitted by
such source or which results in the emission of any air pollutant not previously emitted.

"(5) The term 'owner or operator' means any person who
owns, leases, operates, controls, or supervises a stationary
source.

"(6) The term 'existing source' means any stationary
source other than a new source.

"(b) (1) (A) The Administrator shall, within 90 days after the
date of enactment of the Clean Air Amendments of 1970, publish
(and from time to time thereafter shall revise) a list of categories
of stationary sources. He shall include a category of sources in
such list if he determines it may contribute significantly to air
pollution which causes or contributes to the endangerment of
public health or welfare.

"(B) Within 120 days after the inclusion of a category of
stationary sources in a list under subparagraph (A), the Administrator shall propose regulations, establishing Federal standards
of performance for new sources within such category. The Administrator shall afford interested persons an opportunity for written
comment on such proposed regulations. After considering such
comments, he shall promulgate, within 90 days after such publication, such standards with such modifications as he deems appropriate. The Administrator may, from time to time, revise such
standards following the procedure required by this subsection for
promulgation of such standards. Standards of performance or
revisions thereof shall become effective upon promulgation.

"(2) The Administrator may distinguish among classes, types, and sizes within categories of new sources for the purpose of establishing such standards.

"(3) The Administrator shall, from time to time, issue information on pollution control techniques for categories of new sources and air pollutants subject to the provisions of this section.

"(4) The provisions of this section shall apply to any new source owned or operated by the United States.

"(c) (1) Each State may develop and submit to the Administrator a procedure for implementing and enforcing standards of performance for new sources located in such State. If the Administrator finds the State procedure is adequate, he shall delegate to such State any authority he has under this Act to implement and enforce such standards (except with respect to new sources owned or operated by the United States).

"(2) Nothing in this subsection shall prohibit the Administrator from enforcing any applicable standard of performance under this section.

"(d) (1) The Administrator shall prescribe regulations which shall establish a procedure similar to that provided by section 110 under which each State shall submit to the Administrator a plan which (A) establishes emission standards for any existing source for any air pollutant (i) for which air quality criteria have not been issued or which is not included on a list published under section 108 (a) or 112 (b) (1) (A) but (ii) to which a standard of performance under subsection (b) would apply if such existing source were a new source, and (B) provides for the implementation and enforcement of such emission standards.

"(2) The Administrator shall have the same authority—

"(A) to prescribe a plan for a State in cases where the State fails to submit a satisfactory plan as he would have under section 110 (c) in the case of failure to submit an implementation plan, and

"(B) to enforce the provisions of such plan in cases where the State fails to enforce them as he would have under sections 113 and 114 with respect to an implementation plan.

"(e) After the effective date of standards of performance promulgated under this section, it shall be unlawful for any owner or operator of any new source to operate such source in violation of any standard of performance applicable to such source.

"NATIONAL EMISSION STANDARDS FOR HAZARDOUS AIR POLLUTANTS

"SEC. 112. (a) For purposes of this section—

"(1) The term 'hazardous air pollutant' means an air pollutant to which no ambient air quality standard is applicable and which in the judgment of the Administrator may cause, or contribute to, an increase in mortality or an increase in serious irreversible, or incapacitating reversible, illness.

"(2) The term 'new source' means a stationary source the construction or modification of which is commenced after the Administrator proposes regulations under this section establishing an emission standard which will be applicable to such source.

"(3) The terms 'stationary source,' 'modification,' 'owner or operator' and 'existing source' shall have the same meaning as such terms have under section 111(a).

"(b)(1)(A) The Administrator shall, within 90 days after the date of enactment of the Clean Air Amendments of 1970, publish (and shall from time to time thereafter revise) a list which includes each hazardous air pollutant for which he intends to establish an emission standard under this section.

"(B) Within 180 days after the inclusion of any air pollutant in such list, the Administrator shall publish proposed regulations establishing emission standards for such pollutant together with a notice of a public hearing within thirty days. Not later than 180 days after such publication, the Administrator shall prescribe an emission standard for such pollutant, unless he finds, on the basis of information presented at such hearings, that such pollutant clearly is not a hazardous air pollutant. The Administrator shall establish any such standard at the level which in his judgment provides an ample margin of safety to protect the public health from such hazardous air pollutant.

"(C) Any emission standard established pursuant to this section shall become effective upon promulgation.

"(2) The Administrator shall, from time to time, issue information on pollution control techniques for air pollutants subject to the provisions of this section.

"(c)(1) After the effective date of any emission standard under this section—

"(A) no person may construct any new source or modify any existing source which, in the Administrator's judgment, will emit an air pollutant to which such standard applies unless the Administrator finds that such source if properly operated will not cause emissions in violation of such standard, and

"(B) no air pollutant to which such standard applies may be emitted from any stationary source in violation of such standard, except that in the case of an existing source—

"(i) such standard shall not apply until 90 days after its effective date, and

"(ii) the Administrator may grant a waiver permitting such source a period of up to two years after the effective date of a standard to comply with the standard, if he finds that such period is necessary for the installation of controls and that steps will be taken during the period of the waiver to assure that the health of persons will be protected from imminent endangerment.

"(2) The President may exempt any stationary source from compliance with paragraph (1) for a period of not more than two years if he finds that the technology to implement such standards is not available and the operation of such source is required for reasons of national security. An exemption under this paragraph may be extended for one or more additional periods, each period not to exceed two years. The President shall make a report to Congress with respect to each exemption (or extension thereof) made under this paragraph.

"(d)(1) Each State may develop and submit to the Administrator a procedure for implementing and enforcing *emission standards for hazardous air pollutants* for stationary sources located in such State. If the Administrator finds the State procedure is adequate, he shall delegate to such State any authority he has under this Act to implement and enforce such standards (except with respect to stationary sources owned or operated by the United States).

"(2) Nothing in this subsection shall prohibit the Administrator from enforcing any applicable *emission* standard under this section.

"FEDERAL ENFORCEMENT

"SEC. 113. (a)(1) Whenever, on the basis of any information available to him, the Administrator finds that any person is in violation of any requirement of an applicable implementation plan, the Administrator shall notify the person in violation of the plan and the State in which the plan applies of such finding. If such violation extends beyond the 30th day after the date of the Administrator's notification, the Administrator may issue an order requiring such person to comply with the requirements of such plan or he may bring a civil action in accordance with subsection (b).

"(2) Whenever, on the basis of information available to him, the Administrator finds that violations of an applicable implementation plan are so widespread that such violations appear to result from a failure of the State in which the plan applies to enforce the plan effectively, he shall so notify the State. If the Administrator finds such failure extends beyond the thirtieth day after such notice, he shall give public notice of such finding. During the period beginning with such public notice and ending when such State satisfies the Administrator that it will enforce such plan (hereafter referred to in this section as 'period of Federally assumed enforcement'), the Administrator may enforce any requirement of such plan with respect to any person—

"(A) by issuing an order to comply with such requirement, or

"(B) by bringing a civil action under subsection (b).

"(3) Whenever, on the basis of any information available to him, the Administrator finds that any person is in violation of section 111(e) (relating to new source performance standards) or 112(c) (relating to standards for hazardous emissions), or is in violation of any requirement of section 114 (relating to inspections, etc.), he may issue an order requiring such person to comply with such section or requirement, or he may bring a civil action in accordance with subsection (b).

"(4) An order issued under this subsection (other than an order relating to a violation of section 112) shall not take effect until the person to whom it is issued has had an opportunity to confer with the Administrator concerning the alleged violation. A copy of any order issued under this subsection shall be sent to the State air pollution control agency of any State in which the violation occurs. Any order issued under this subsection shall state with reasonable specificity the nature of the violation, specify a time for compliance which the Administrator determines is reasonable, taking into account the seriousness of the violation and any good faith efforts to comply with applicable requirements. In any case in which an order under this subsection (or notice to a violator under paragraph (1)) is issued to a corporation, a copy of such order (or notice) shall be issued to appropriate corporate officers.

"(b) The Administrator may commence a civil action for appropriate relief, including a permanent or temporary injunction, whenever any person—

"(1) violates or fails or refuses to comply with any order issued under subsection (a); or

"(2) violates any requirement of an applicable implementation plan during any period of Federally assumed enforcement more than 30 days after having been notified by the Administrator under subsection (a)(1) of a finding that such person is violating such requirement; or

"(3) violates section 111(e) or 112(c); or

"(4) fails or refuses to comply with any requirement of section 114.

Any action under this subsection may be brought in the district court of the United States for the district in which the defendant is located or resides or is doing business, and such court shall have jurisdiction to restrain such violation and to require compliance. Notice of the commencement of such action shall be given to the appropriate State air pollution control agency.

"(c) (1) Any person who knowingly—

"(A) violates any requirement of an applicable implementation plan during any period of Federally assumed enforcement more than 30 days after having been notified by the Administrator under subsection (a)(1) that such person is violating such requirement, or

"(B) violates or fails or refuses to comply with any order issued by the Administrator under subsection (a), or

"(C) violates section 111(e) or section 112(c)

shall be punished by a fine of not more than $25,000 per day of violation, or by imprisonment for not more than one year, or by both. If the conviction is for a violation committed after the first conviction of such person under this paragraph, punishment shall be by a fine of not more than $50,000 per day of violation, or by imprisonment for not more than two years, or by both.

"(2) Any person who knowingly makes any false statement, representation, or certification in any application, record, report, plan, or other document filed or required to be maintained under this Act or who falsifies, tampers with, or knowingly renders inaccurate any monitoring device or method required to be maintained under this Act, shall upon conviction, be punished by a fine of not more than $10,000, or by imprisonment for not more than six months, or by both.

"INSPECTIONS, MONITORING, AND ENTRY

"SEC. 114. (a) For the purpose (i) of developing or assisting in the development of any implementation plan under section 110 or 111(d), any standard of performance under section 111, or any emission standard under section 112 (ii) of determining whether any person is in violation of any such standard or any requirement of such a plan, or (iii) carrying out section 303—

"(1) the Administrator may require the owner or operator of any emission source to (A) establish and maintain such records, (B) make such reports, (C) install, use, and maintain such monitoring equipment or methods, (D) sample such emissions (in accordance with such methods, at such locations, at such intervals, and in such manner as the Administrator shall prescribe), and (E) provide such other information, as he may reasonably require; and

"(2) the Administrator of his authorized representative, upon presentation of his credentials—

"(A) shall have a right of entry to, upon, or through any premises in which an emission source is located or in which any records required to be maintained under paragraph (1) of this section are located, and

"(B) may at reasonable times have access to and copy any records, inspect any monitoring equipment or method required under paragraph (1), and sample any emissions which the owner or operator of such source is required to sample under paragraph (1).

"(b)(1) Each State may develop and submit to the Administrator a procedure for carrying out this section in such State. If the Administrator finds the State procedure is adequate, he may delegate to such State any authority he has to carry out this section (except with respect to new sources owned or operated by the United States).

"(2) Nothing in this subsection shall prohibit the Administrator from carrying out this section in a State.

"(c) Any records, reports or information obtained under subsection (a) shall be available to the public, except that upon a showing satisfactory to the Administrator by any person that records, reports, or information, or particular part thereof, (other than emission data) to which the Administrator has access under this section if made public, would divulge methods or processes entitled to protection as trade secrets of such person, the Administrator shall consider such record, report, or information or particular portion thereof confidential in accrdance with the purposes of section 1905 of title 18 of the United States Code, except that such record, report, or information may be disclosed to other officers, employees, or authorized representatives of the United

States concerned with carrying out this Act or when relevant in any proceeding under this Act.

"ABATEMENT BY MEANS OF CONFERENCE PROCEDURE
IN CERTAIN CASES

"SEC. 115. (a) The pollution of the air in any State or States which endangers the health or welfare of any persons and which is covered by subsection (b) or (c) shall be subject to abatement as provided in this section.

"(b) (1) Whenever requested by the Governor of any State, a State air pollution control agency, or (with the concurrence of the Governor and the State air pollution control agency for the State in which the municipality is situated) the governing body of any municipality, the Administrator shall, if such request refers to air pollution which is alleged to endanger the health or welfare of persons in a State other than that in which the discharge or discharges (causing or contributing to such pollution) originate, give formal notification thereof to the air pollution control agency of the municipality where such discharge or discharges originate, to the air pollution control agency of the State in which such municipality is located, and to the interstate air pollution control agency, if any, in whose jurisdictional area such municipality is located, and shall call promptly a conference of such agency or agencies and of the air pollution control agencies of the municipalities which may be adversely affected by such pollution, and the air pollution control agency, if any, of each State, or for each area, in which any such municipality is located.

"(2) Whenever requested by the Governor of any State, a State air pollution control agency, or (with the concurrence of the Governor and the State air pollution control agency for the State in which the municipality is situated) the governing body of any municipality, the Administrator shall, if such request refers to alleged air pollution which is endangering the health or welfare of persons only in the State in which the discharge or discharges (causing or contributing to such pollution) originate and if a municipality affected by such air pollution, or the municipality in which such pollution originates, has either made or concurred in such request, give formal notification thereof to the State air pollution control agency, to the air pollution control agencies of the municipality where such discharge or discharges originate, and of the municipality or municipalities alleged to be adversely affected thereby, and to any interstate air pollution control agency, whose jurisdictional area includes any such municipality and shall promptly call a conference of such agency or agencies, unless in the judgment of the Administrator, the effect of such pollution is not of such significance as to warrant exercise of Federal jurisdiction under this section.

"(3) The Administrator may, after consultation with State officials of all affected States, also call such a conference whenever, on the basis of reports, surveys, or studies, he has reason to believe that any pollution referred to in subsection (a) is occurring and is endangering the health and welfare of persons in a State other than that in which the discharge or discharges originate. The Administrator shall invite the cooperation of any municipal, State, or interstate air pollution control agencies having jurisdiction in the affected area on any surveys or studies forming the basis of conference action.

"(4) A conference may not be called under this subsection with respect to an air pollutant for which (at the time the conference is called) a national primary or secondary ambient air quality standard is in effect under section 109.

"(c) Whenever the Administrator, upon receipt of reports, surveys, or studies from any duly constituted international agency, has reason to believe that any pollution referred to in subsection (a) which endangers the health or welfare of persons in a foreign country is occurring, or whenever the Secretary of State requests him to do so with respect to such pollution which the Secretary of State alleges is of such a nature, the Administrator shall give formal notification thereof to the air pollution control agency of the municipality where such discharge or discharges originate, to the air pollution control agency of the State in which such municipality is located, and to the interstate air pollution control agency, if any, in the jurisdictional area of which such municipality is located, and shall call promptly a conference of such agency or agencies. The Administrator shall invite the foreign country which may be adversely affected by the pollution to attend and participate in the conference, and the representative of such country shall, for the purpose of the conference and any further proceeding resulting from such conference, have all the rights of a State air pollution control agency. This subsection shall apply only to a foreign country which the Administrator determines has given the United States essentially the same rights with respect to the prevention or control of air pollution occurring in that country as is given that country by this subsection.

"(d) (1) The agencies called to attend any conference under this section may bring such persons as they desire to the conference. The Administrator shall deliver to such agencies and make available to other interested parties, at least thirty days prior to any such conference, a Federal report with respect to the matters before the conference, including data and conclusions or findings (if any) ; and shall give at least thirty days' prior notice of the conference date to any such agency, and to the public by publication on at least three different days in a newspaper or newspapers of general circulation in the area. The chairman of the conference shall give interested parties an opportunity to present their views to the conference with respect to such Federal report, conclusions or findings (if any), and other pertinent information. The Administrator shall provide that a transcript be maintained of the proceedings of the conference and that a copy of such transcript be made available on request of any participant in the conference at the expense of such participant.

"(2) Following this conference, the Administrator shall prepare and forward to all air pollution control agencies attending the conference a summary of conference discussions including (A) occurrence of air pollution subject to abatement under this Act; (B) adequacy of measures taken toward abatement of the pollution; and (C) nature of delays, if any, being encountered in abating the pollution.

"(e) If the Administrator believes, upon the conclusion of the conference or thereafter, that effective progress toward abatement of such pollution is not being made and that the health or welfare of any persons is being endangered, he shall recommend to the appropriate State, interstate, or municipal air pollution control agency (or to all such agencies) that the necessary remedial action be taken. The Administrator shall allow at least six months from the date he makes such recommendations for the taking of such recommended action.

"(f) (1) If, at the conclusion of the period so allowed, such remedial action or other action which in the judgment of the Administrator is reasonably calculated to secure abatement of such pollution has not been taken, the Administrator shall call a public hearing, to be held in or near one or more of the places where the discharge or discharges causing or contributing to such pollution originated, before a hearing board of five or more persons appointed by the Administrator. Each State in which any discharge causing or contributing to such pollution originates and each State claiming to be adversely affected by such pollution shall be given an opportunity to select one member of such hearing board and each Federal department, agency, or instrumentality having a substantial interest in the subject matter as determined by the Administrator shall be given an opportunity to select one member of such hearing board, and one member shall be a representative of the appropriate interstate air pollution agency if one exists, and not less than a majority of such hearing board shall be persons other than officers or employees of the Environmental Protection Agency. At least three weeks' prior notice of such hearing shall be given to the State, interstate, and municipal air pollution control agencies called to attend such hearing and to the alleged polluter or polluters. All interested parties shall be given a reasonable opportunity to present evidence to such hearing board.

"(2) On the basis of evidence presented at such hearing, the hearing board shall make findings as to whether pollution referred to in subsection (a) is occurring and whether effective progress toward abatement thereof is being made. If the hearing board finds such pollution is occurring and effective progress toward abatement thereof is not being made it shall make recommendations to the Administrator concerning the measures, if any, which it finds to be reasonable and suitable to secure abatement of such pollution.

"(3) The Administrator shall send such findings and recommendations to the person or persons discharging any matter causing or contributing to such pollution; to air pollution control agencies of the State or States and of the municipality or munici-

43

palities where such discharge or discharges originate; and to any interstate air pollution control agency whose jurisdictional area includes any such municipality, together with a notice specifying a reasonable time (not less than six months) to secure abatement of such pollution.

"(g) If action reasonably calculated to secure abatement of the pollution within the time specified in the notice following the public hearing is not taken the Administrator—

(1) in the case of pollution of air which is endangering the health or welfare of persons (A) in a State other than that in which the discharge or discharges (causing or contributing to such pollution) originate, or (B) in a foreign country which has participated in a conference called under subsection (c) of this section and in all proceedings under this section resulting from such conference, may request the Attorney General to bring a suit on behalf of the United States in the appropriate United States district court to secure abatement of the pollution.

"(2) in the case of pollution of air which is endangering the health or welfare of persons only in the State in which the discharge or discharges (causing or contributing to such pollution) originate, at the request of the Governor of such State, shall provide such technical and other assistance as in his judgment is necessary to assist the State in judicial proceedings to secure abatement of the pollution under State or local law or, at the request of the Governor of such State, shall request the Attorney General to bring suit on behalf of the United States in the appropriate United States district court to secure abatement of the pollution.

"(h) The court shall receive in evidence in any suit brought in a United States court under subsection (g) of this section a transcript of the proceedings before the board and a copy of the board's recommendations and shall receive such further evidence as the court in its discretion deems proper. The court, giving due consideration to the practicability of complying with such standards as may be applicable and to the physical and economic feasibility of securing abatement of any pollution proved, shall have jurisdiction to enter such judgment, and orders enforcing such judgment, as the public interest and the equities of the case may require.

"(i) Members of any hearing board appointed pursuant to subsection (f) who are not regular full-time officers or employees of the United States shall, while participating in the hearing conducted by such board or otherwise engaged on the work of such board, be entitled to receive compensation at a rate fixed by the Administration, but not exceeding $100 per diem, including traveltime, and while away from their homes or regular places of business they may be allowed travel expenses, including per diem in lieu of subsistence, as authorized by law (5 U.S.C. 73b–2) for persons in the Government service employed intermittently.

"(j) (1) In connection with any conference called under this section, the Administrator is authorized to require any person whose activities result in the emission of air pollutants causing or contributing to air pollution to file with him, in such form as he may prescribe, a report, based on existing data, furnishing to the Administrator such information as may reasonably be required as to the character, kind, and quantity of pollutants discharged and the use of devices or other means to prevent or reduce the emission of pollutants by the person filing such a report. After a conference has been held with respect to any such pollution the Administrator shall require such reports from the person whose activities result in such pollution only to the extent recommended by such conference. Such report shall be made under oath or otherwise, as the Administrator may prescribe, and shall be filed with the Administrator within such reasonable period as the Administrator may prescribe, unless additional time be granted by the Administrator. No person shall be required in such report to divulge trade secrets or secret processes and all information reported shall be considered confidential for the purposes of section 1905 of title 18 of the United States Code.

"(2) If any person required to file any report under this subsection shall fail to do so within the time fixed by the Administrator for filing the same, and such failure shall continue for thirty days after notice of such default, such person shall forfeit to the United States the sum of $100 for each and every day of the continuance of such failure, which forfeiture shall be payable into the Treasury of the United States, and shall be recoverable in a civil suit in the name of the United States brought in the district where such person has his principal office or in any district in which he does business: *Provided,* That the Administrator may upon application therefor remit or mitigate any forfeiture provided for under this subsection and he shall have authority to determine the facts upon all such applications.

"(3) It shall be the duty of the various United States attorneys, under the direction of the Attorney General of the United States, to prosecute for the recovery of such forfeitures.

"(k) No order or judgment under this section, or settlement, compromise, or agreement respecting any action under this section (whether or not entered or made before the date of enactment of the Clean Air Amendments of 1970) shall relieve any person of any obligation to comply with any requirement of an applicable implementation plan, or with any standard prescribed under section 111 or 112.

"RETENTION OF STATE AUTHORITY

"SEC. 116. Except as otherwise provided in sections 209, 211(c)(4), and 233 (preempting certain State regulation of moving sources) nothing in this Act shall preclude or deny the right of any State or political subdivision thereof to adopt or enforce (1) any standard or limitation respecting emissions of air pollutants or (2) any requirement respecting control or abatement of air pollution; except that if an emission standard or limitation is in effect under an applicable implementation plan or under section

45

111 or 112, such State or political subdivision may not adopt or enforce any emission standard or limitation which is less stringent than the standard or limitation under such plan or section.

"SEC. 117. (a) (1) There is hereby established in the Environmental Protection Agency an Air Quality Advisory Board, composed of the Administrator or his designee, who shall be Chairman, and fifteen members appointed by the President, none of whom shall be Federal officers or employees. The appointed members, having due regard for the purposes of this Act, shall be selected from among representatives of various State, interstate, and local governmental agencies, of public or private interests contributing to, affected by, or concerned with air pollution, and of other public and private agencies, organizations, or groups demonstrating an active interest in the field of air pollution prevention and control, as well as other individuals who are expert in this field.

"(2) Each member appointed by the President shall hold office for a term of three years, except that (A) any member appointed to fill a vacany occurring prior to the expiriation of the term for which his predecessor was appointed shall be appointed for the remainder of such term, and (B) the terms of office of the members first taking office pursuant to this subsection shall expire as follows: five at the end of one year after the date of appointment, five at the end of two years after such date, and five at the end of three years after such date, as designated by the President at the time of appointment, and (C) the term of any member under the preceding provisions shall be extended until the date on which his successor's appointment is effective. None of the members shall be eligible for reappointment within one year after the end of his preceding term, unless such term was for less than three years.

"(b) The Board shall advise and consult with the Administrator on matters of policy relating to the activities and functions of the Administrator under this Act and make such recommendations as it deems necessary to the President.

"(c) Such clerical and technical assistance as may be necessary to discharge the duties of the Board and such other advisory committees as hereinafter authorized shall be provided from the personnel of the Environmental Protection Agency.

"(d) In order to obtain assistance in the development and implementation of the purposes of this Act including air quality criteria, recommended control techniques, standards, research and development, and to encourage the continued efforts on the part of industry to improve air quality and to develop economically feasible methods for the control and abatement of air pollution, the Administrator shall from time to time establish advisory committees. Committee members shall include, but not be limited to, persons who are knowledgeable concerning air quality from the standpoint of health, welfare, economics, or technology.

"(e) The members of the Board and other advisory committees appointed pursuant to this Act who are not officers or employees of the United States while attending conferences or meetings of the Board or while otherwise serving at the request of the Administrator, shall be entitled to receive compensation at a rate to be fixed by the Administrator, but not exceeding $100 per diem, including traveltime, and while away from their homes or regular places of business they may be allowed travel expenses, including per diem in lieu of subsistence, as authorized by section 5703 of title 5 of the United States Code for persons in the Government service employed intermittently.

"(f) Prior to—

"(1) issuing criteria for an air pollutant under section 108(a)(2),

"(2) publishing any list under section 111(b)(1)(A) or 112(b)(1)(A),

"(3) publishing any standard under section 111(b)(1)(B) or section 112(b)(1)(B), or

"(4) publishing any regulation under section 202(a),

the Administrator shall, to the maximum extent practicable within the time provided, consult with appropriate advisory committees, independent experts, and Federal departments and agencies.

"CONTROL OF POLLUTION FROM FEDERAL FACILITIES

"SEC. 118. Each department, agency, and instrumentality of the executive, legislative, and judicial branches of the Federal Government (1) having jurisdiction over any property or facility, or (2) engaged in any activity resulting, or which may result, in the discharge of air pollutants, shall comply with Federal, State, interstate, and local requirements respecting control and abatement of air pollution to the same extent that any person is subject to such requirements. The President may exempt any emission source of any department, agency, or instrumentality in the executive branch from compliance with such a requirement if he determines it to be in the paramount interest of the United States to do so, except that no exemption may be granted from section 111, and an exemption from section 112 may be granted only in accordance with section 112(c). No such exemption shall be granted due to lack of appropriation unless the President shall have specifically requested such appropriation as a part of the budgetary process and the Congress shall have failed to make available such requested appropriation. Any exemption shall be for a period not in excess of one year, but additional exemptions may be granted for periods of not to exceed one year upon the President's making a new determination. The President shall report each January to the Congress all exemptions from the requirements of this section granted during the preceding calendar year, together with his reason for granting each such exemption.

"TITLE II—EMISSION STANDARDS FOR MOVING SOURCES

"SEC. 201. This title may be cited as the 'National Emission Standards Act.'

"PART A—MOTOR VEHICLE EMISSION AND FUEL STANDARDS

"ESTABLISHMENT OF STANDARDS

"SEC. 202. (a) Except as otherwise provided in subsection (b)—

"(1) The Administrator shall by regulation prescribe (and from time to time revise) in accordance with the provisions of this section, standards applicable to the emission of any air pollutant from any class or classes of new motor vehicles or new motor vehicle engines, which in his judgment causes or contributes to, or is likely to cause or to contribute to, air pollution which endangers the public health or welfare. Such standards shall be applicable to such vehicles and engines for their useful life (as determined under subsection (d)), whether such vehicles and engines are designed as complete systems or incorporate devices to prevent or control such pollution.

"(2) Any regulation prescribed under this subsection (and any revision thereof) shall take effect after such period as the Administrator finds necessary to permit the development and application of the requisite technology, giving appropriate consideration to the cost of compliance within such period.

"(b) (1) (A) The regulations under subsection (a) applicable to emissions of carbon monoxide and hydrocarbons from light duty vehicles and engines manufactured during or after model year 1975 shall contain standards which require a reduction of at least 90 per centum from emissions of carbon monoxide and hydrocarbons allowable under the standards under this section applicable to light duty vehicles and engines manufactured in model year 1970.

"(B) The regulations under subsection (a) applicable to emissions of oxides of nitrogen from light duty vehicles and engines manufactured during or after model year 1976 shall contain standards which require a reduction of at least 90 per centum from the average of emissions of oxides of nitrogen actually measured from light duty vehicles manufactured during model year 1971 which are not subject to any Federal or State emission standard for oxides of nitrogen. Such average of emissions shall be determined by the Administrator on the basis of measurements made by him.

"(2) Emission standards under paragraph (1), and measurement techniques on which such standards are based (if not promulgated prior to the date of enactment of the Clean Air Amendments of 1970), shall be prescribed by regulation within 180 days after such date.

"(3) For purposes of this part—

"(A) (i) The term 'model year' with reference to any specific calendar year means the manufacturer's annual production period (as determined by the Administrator) which includes January 1 of such calendar year. If the manufacturer has no annual production period, the term 'model year' shall mean the calendar year.

"(ii) For the purpose of assuring that vehicles and engines manufactured before the beginning of a model year were not manufactured for purposes of circumventing the effective date of a standard required to be prescribed by subsection (b), the Administrator may prescribe regulations defining 'model year' otherwise than as provided in clause (i).

"(B) The term 'light duty vehicles and engines' means new light duty motor vehicles and new light duty motor vehicle engines, as determined under regulations of the Administrator.

"(4) On July 1 of 1971, and of each year thereafter, the Administrator shall report to the Congress with respect to the development of systems necessary to implement the emission standards established pursuant to this section. Such reports shall include information regarding the continuing effects of such air pollutants subject to standards under this section on the public health and welfare, the extent and progress of efforts being made to develop the necessary systems, the costs associated with development and application of such systems, and following such hearings as he may deem advisable, any recommendations for additional congressional action necessary to achieve the purposes of this Act. In gathering information for the purposes of this paragraph and in connection with any hearing, the provisions of section 307(a) (relating to subpenas) shall apply.

"(5) (A) At any time after January 1, 1972, any manufacturer may file with the Administrator an application requesting the suspension for one year only of the effective date of any emission standard required by paragraph (1) (A) with respect to such manufacturer. The Administrator shall make his determination with respect to any such application within 60 days. If he determines, in accordance with the provisions of this subsection, that such suspension should be granted, he shall simultaneously with such determination prescribe by regulation interim emission standards which shall apply (in lieu of the standards required to be prescribed, by paragraph (1) (A)) to emissions of carbon monoxide or hydrocarbons (or both) from such vehicles and engines manufactured during model year 1975.

"(B) At any time after January 1, 1973, any manufacturer may file with the Administrator an application requesting the suspension, for one year only of the effective date of any emission standard required by paragraph (1) (B) with respect to such manufacturer. The Administrator shall make his determination

49

with respect to any such application within 60 days. If he determines, in accordance with the provisions of this subsection, that such suspension should be granted, he shall simultaneously with such determination prescribe by regulation interim emission standards which shall apply (in lieu of the standards required to be prescribed by paragraph (1)(B)) to emissions of oxides of nitrogen from such vehicles and engines manufactured during model year 1976.

"(C) Any interim standards prescribed under this paragraph shall reflect the greatest degree of emission control which is achievable by application of technology which the Administrator determines is available, giving appropriate consideration to the cost of applying such technology within the period of time available to manufacturers.

"(D) Within 60 days after receipt of the application for any such suspension, and after public hearing, the Administrator shall issue a decision granting or refusing such suspension. The Administrator shall grant such suspension only if he determines that (i) such suspension is essential to the public interest or the public health and welfare of the United States, (ii) all good faith efforts have been made to meet the standards established by this subsection, (iii) the applicant has established that effective control technology, processes, operating methods, or other alternatives are not available or have not been available for a sufficient period of time to achieve compliance prior to the effective date of such standards, and (iv) the study and investigation of the National Academy of Sciences conducted pursuant to subsection (c) and other information available to him has not indicated that technology, processes, or other alternatives are available to meet such standards.

"(E) Nothing in this paragraph shall extend the effective date of any emission standard required to be prescribed under this subsection for more than one year.

"(c)(1) The Administrator shall undertake to enter into appropriate arrangements with the National Academy of Sciences to conduct a comprehensive study and investigation of the technological feasibility of meeting the emissions standards required to be prescribed by the Administrator by subsection (b) of this section.

"(2) Of the funds authorized to be appropriated to the Administrator by this Act, such amounts as are required shall be available to carry out the study and investigation authorized by paragraph (1) of this subsection.

"(3) In entering into any arrangement with the National Academy of Sciences for conducting the study and investigation authorized by paragraph (1) of this subsection, the Administrator shall request the National Academy of Sciences to submit semiannual reports on the progress of its study and investigation to the Administrator and the Congress, beginning not later than July 1, 1971, and continuing until such study and investigation is completed.

"(4) The Administrator shall furnish to such Academy at its request any information which the Academy deems necessary for

50

the purpose of conducting the investigation and study authorized by paragraph (1) of this subsection. For the purpose of furnishing such information, the Administrator may use any authority he has under this Act (A) to obtain information from any person, and (B) to require such person to conduct such tests, keep such records, and make such reports respecting research or other activities conducted by such person as may be reasonably necessary to carry out this subsection:

"(d) The Administrator shall prescribe regulations under which the useful life of vehicles and engines shall be determined for purposes of subsection (a)(1) of this section and section 207. Such regulations shall provide that useful life shall—

"(1) in the case of light duty vehicles and light duty vehicle engines, be a period of use of five years or of fifty thousand miles (or the equivalent), whichever first occurs; and

"(2) in the case of any other motor vehicle or motor vehicle engine, be a period of use set forth in paragraph (1) unless the Administrator determines that a period of use of greater duration or mileage is appropriate.

"(e) In the event a new power source or propulsion system for new motor vehicles or new motor vehicle engines is submitted for certification pursuant to section 206(a), the Administrator may postpone certification until he has prescribed standards for any air pollutants emitted by such vehicle or engine which cause or contribute to, or are likely to cause or contribute to, air pollution which endangers the public health or welfare but for which standards have not been prescribed under subsection (a).

"PROHIBITED ACTS

"SEC. 203. (a) The following acts and the causing thereof are prohibited—

"(1) in the case of a manufacturer of new motor vehicles or new motor vehicle engines for distribution in commerce, the sale, or the offering for sale, or the introduction, or delivery for introduction, into commerce, or (in the case of any person, except as provided by regulation of the Administrator), the importation into the United States, of any new motor vehicle or new motor vehicle engine, manufactured after the effective date of regulations under this part which are applicable to such vehicle or engine unless such vehicle or engine is covered by a certificate of conformity issued (and in effect) under regulations prescribed under this part (except as provided in subsection (b));

"(2) for any person to fail or refuse to permit access to or copying of records or to fail to make reports or provide information, required under section 208;

"(3) for any person to remove or render inoperative any device or element of design installed on or in a motor vehicle or motor vehicle engine in compliance with regulations under this title prior to its sale and delivery to the ultimate purchaser, or for any manufacturer or dealer knowingly to remove or render inoperative any such device or element of design after such sale and delivery to the ultimate purchaser; or

"(4) for any manufacturer of a new motor vehicle or new motor vehicle engine subject to standards prescribed under section 202—

"(A) to sell or lease any such vehicle or engine unless such manufacturer has complied with the requirements of section 207(a) and (b) with respect to such vehicle or engine, and unless a label or tag is affixed to such vehicle or engine in accordance with section 207(c)(3), or

"(B) to fail or refuse to comply with the requirements of section 207(c) or (e).

"(b)(1) The Administrator may exempt any new motor vehicle or new motor vehicle engine from subsection (a), upon such terms and conditions as he may find necessary for the purpose of research, investigations, studies, demonstrations, or training, or for reasons of national security.

"(2) A new motor vehicle or new motor vehicle engine offered for importation or imported by any person in violation of subsection (a) shall be refused admission into the United States, but the Secretary of the Treasury and the Administrator may, by joint regulation, provide for deferring final determination as to admission and authorizing the delivery of such a motor vehicle or engine offered for import to the owner or consignee thereof upon such terms and conditions (including the furnishing of a bond) as may appear to them appropriate to insure that any such motor vehicle or engine will be brought into conformity with the standards, requirements, and limitations applicable to it under this part. The Secretary of the Treasury shall, if a motor vehicle or engine is finally refused admission under this paragraph, cause disposition thereof in accordance with the customs laws unless it is exported, under regulations prescribed by such Secretary, within ninety days of the date of notice of such refusal or such additional time as may be permitted pursuant to such regulations, except that disposition in accordance with the customs laws may not be made in such manner as may result, directly or indirectly, in the sale, to the ultimate consumer, of a new motor vehicle or new motor vehicle engine that fails to comply with applicable standards of the Administrator under this part.

"(3) A new motor vehicle or new motor vehicle engine intended solely for export, and so labeled or tagged on the outside of the container and on the vehicle or engine itself, shall be subject to the provisions of subsection (a), except that if the country of export has emission standards which differ from the standards prescribed under subsection (a), then such vehicle or engine shall comply with the standards of such country of export.

"(c) Upon application therefor, the Administrator may exempt from section 203(a)(3) any vehicles (or class thereof) manufactured before the 1974 model year from section 203(a)(3) for the purpose of permitting modifications to the emission control device or system of such vehicle in order to use fuels other than those specified in certification testing under section 206(a)(1), if the Administrator, on the basis of information submitted by the applicant, finds that such modification will not result in such vehicle or engine not complying with standards under section 202 applicable to such vehicle or engine. Any such exemption shall identify (1) the vehicle or vehicles so exempted, (2) the specific nature of the modification, and (3) the person or class of persons to whom the exemption shall apply.

"INJUNCTION PROCEEDINGS

"SEC. 204. (a) The district courts of the United States shall have jurisdiction to restrain violations of paragraph (1), (2), (3), or (4) of section 203(a).

"(b) Actions to restrain such violations shall be brought by and in the name of the United States. In any such action, subpenas for witnesses who are required to attend a district court in any district may run into any other district.

"PENALTIES

"SEC. 205. Any person who violates paragraph (1), (2), (3), or (4) of section 203(a) shall be subject to a civil penalty of not more than $10,000. Any such violation with respect to paragraph (1), (2), or (4) of section 203(a) shall constitute a separate offense with respect to each motor vehicle or motor vehicle engine.

"MOTOR VEHICLE AND MOTOR VEHICLE ENGINE COMPLIANCE
TESTING AND CERTIFICATION

"SEC. 206. (a)(1) The Administrator shall test, or require to be tested in such manner as he deems appropriate, any new motor vehicle or new motor vehicle engine submitted by a manufacturer to determine whether such vehicle or engine conforms with the regulations prescribed under section 202 of this Act. If such vehicle or engine conforms to such regulations, the Administrator shall issue a certificate of conformity upon such terms, and for such period (not in excess of one year), as he may prescribe.

"(2) The Administrator shall test any emission control system incorporated in a motor vehicle or motor vehicle engine submitted to him by any person, in order to determine whether such system enables such vehicle or engine to conform to the standards required to be prescribed under section 202(b) of this Act. If the Administrator finds on the basis of such tests that such vehicle or engine conforms to such standards, the Administrator shall issue a verification of compliance with emission standards for such system when incorporated in vehicles of a class of which the tested vehicle is representative. He shall inform manufacturers and the National Academy of Sciences, and make available to the public, the results of such tests. Tests under this paragraph shall be conducted under such terms and conditions (including requirements for preliminary testing by qualified independent laboratories) as the Administrator may prescribe by regulations.

"(b) (1) In order to determine whether new motor vehicles or new motor vehicle engines being manufactured by a manufacturer do in fact conform with the regulations with respect to which the certificate of conformity was issued, the Administrator is authorized to test such vehicles or engines. Such tests may be conducted by the Administrator directly or, in accordance with conditions specified by the Administrator, by the manufacturer.

"(2) (A) (i) If, based on tests conducted under paragraph (1) on a sample of new vehicles or engines covered by a certificate of conformity, the Administrator determines that all or part of the vehicles or engines so covered do not conform with the regulations with respect to which the certificate of conformity was issued, he may suspend or revoke such certificate in whole or in part, and shall so notify the manufacturer. Such suspension or revocation shall apply in the case of any new motor vehicles or new motor vehicle engines manufactured after the date of such notification (or manufactured before such date if still in the hands of the manufacturer), and shall apply until such time as the Administrator finds that vehicles and engines manufactured by the manufacturer do conform to such regulations. If, during any period of suspension or revocation, the Administrator finds that a vehicle or engine actually conforms to such regulations, he shall issue a certificate of conformity applicable to such vehicle or engine.

"(ii) If, based on tests conducted under paragraph (1) on any new vehicle or engine, the Administrator determines that such vehicle or engine does not conform with such regulations, he may suspend or revoke such certificate insofar as it applies to such vehicle or engine until such time as he finds such vehicle or engine actually so conforms with such regulations, and he shall so notify the manufacturer.

"(B) (i) At the request of any manufacturer the Administrator shall grant such manufacturer a hearing as to whether the tests have been properly conducted or any sampling methods have been properly applied, and make a determination on the record with respect to any suspension or revocation under subparagraph (A); but suspension or revocation under subparagraph (A) shall not be stayed by reason of such hearing.

"(ii) In any case of actual controversy as to the validity of any determination under clause (i), the manufacturer may at any time prior to the 60th day after such determination is made file a petition with the United States court of appeals for the circuit wherein such manufacturer resides or has his principal place of business for a judicial review of such determination. A copy of the petition shall be forthwith transmitted by the clerk of the court to the Administrator or other officer designated by him for that purpose. The Administrator thereupon shall file in the court the record of the proceedings on which the Administrator based his determination, as provided in section 2112 of title 28 of the United States Code.

"(iii) If the petitioner applies to the court for leave to adduce additional evidence, and shows to the satisfaction of the court that such additional evidence is material and that there were reasonable grounds for the failure to adduce such evidence in the proceeding before the Administrator, the court may order such additional evidence (and evidence in rebuttal thereof) to be taken before the Administrator, in such manner and upon such terms and conditions as the court may deem proper. The Administrator may modify his findings as to the facts, or make new findings, by reason of the additional evidence so taken and he shall file such modified or new findings, and his recommendation, if any, for the modification or setting aside of his original determination, with the return of such additional evidence.

"(iv) Upon the filing of the petition referred to in clause (ii), the court shall have jurisdiction to review the order in accordance with chapter 7 of title 5, United States Code, and to grant appropriate relief as provided in such chapter.

"(c) For purposes of enforcement of this section, officers or employees duly designated by the Administrator, upon presenting appropriate credentials to the manufacturer or person in charge, are authorized (1) to enter, at reasonable times, any plant or other establishment of such manufacturers, for the purpose of conducting tests of vehicles or engines in the hands of the manufacturer, or (2) to inspect at reasonable times, records, files, papers, processes, controls, and facilities used by such manufacturer in conducting tests under regulations of the Administrator. Each such inspection shall be commenced and completed with reasonable promptness.

"(d) The Administrator shall by regulation establish methods and procedures for making tests under this section.

"(e) The Administrator shall announce in the Federal Register and make available to the public the results of his tests of any motor vehicle or motor vehicle engine submitted by a manufacturer under subsection (a) as promptly as possible after the enactment of the Clean Air Amendments of 1970 and at the beginning of each model year which begins thereafter. Such results shall be described in such nontechnical manner as will reasonably disclose to prospective ultimate purchasers of new motor vehicles and new motor vehicle engines the comparative performance of the vehicles and engines tested in meeting the standards prescribed under section 202 of this Act.

"COMPLIANCE BY VEHICLES AND ENGINES IN ACTUAL USE

"SEC. 207. (a) Effective with respect to vehicles and engines manufactured in model years beginning more than 60 days after the date of the enactment of the Clean Air Amendments of 1970, the manufacturer of each new motor vehicle and new motor vehicle engine shall warrant to the ultimate purchaser and each subsequent purchaser that such vehicle or engine is (1) designed, built, and equipped so as to conform at the time of sale with applicable regulations under section 202, and (2) free from defects in materials and workmanship which cause such vehicle or engine to fail to conform with applicable regulations for its useful life (as determined under sec. 202(d)) .

"(b) If the Administrator determines that (i) there are available testing methods and procedures to ascertain whether, when in actual use throughout its useful life (as determined under section 202(d)), each vehicle and engine to which regulations under section 202 apply complies with the emission standards of such regulations, (ii) such methods and procedures are in accordance with good engineering practices, and (iii) such methods and procedures are reasonably capable of being correlated with tests conducted under section 206(a)(1), then—

"(1) he shall establish such methods and procedures by regulation, and

"(2) at such time as he determines that inspection facilities or equipment are available for purposes of carrying out testing methods and procedures established under paragraph (1), he shall prescribe regulations which shall require manufacturers to warrant the emission control device or system of each new motor vehicle or new motor vehicle engine to which a regulation under section 202 applies and which is manufactured in a model year beginning after the Administrator first prescribes warranty regulations under this paragraph (2). The warranty under such regulations shall run to the ultimate purchaser and each subsequent purchaser and shall provide that if—

"(A) the vehicle or engine is maintained and operated in accordance with instructions under subsection (c)(3),

"(B) it fails to conform at any time during its useful life (as determined under section 202(d)) to the regulations prescribed under section 202, and

"(C) such nonconformity results in the ultimate purchaser (or any subsequent purchaser) of such vehicle or engine having to bear any penalty or other sanction (including the denial of the right to use such vehicle or engine) under State or Federal law,

then such manufacturer shall remedy such nonconformity under such warranty with the cost thereof to be borne by the manufacturer.

"(c) Effective with respect to vehicles and engines manufactured during model years beginning more than 60 days after the date of enactment of the Clean Air Amendments of 1970—

"(1) If the Administrator determines that a substantial number of any class or category of vehicles or engines, although properly maintained and used, do not conform to the regulations prescribed under section 202, when in actual use throughout their useful life (as determined under section 202(d)), he shall immediately notify the manufacturer thereof of such nonconformity, and he shall require the manufacturer to submit a plan for remedying the nonconformity of the vehicles or engines with respect to which such notification is given. The plan shall provide that the nonconformity of any such vehicles or engines which are properly used and maintained will be remedied at the expense of the manufacturer. If the manufacturer disagrees with such determination of nonconformity and so advises the Administrator, the Administrator shall afford the manufacturer and other interested persons an opportunity to present their views and evidence in support thereof at a public hearing. Unless, as a result of such hearing the Administrator withdraws such determination of nonconformity, he shall ,within 60 days after the completion of such hearing, order the manufacturer to provide prompt notification of such nonconformity in accordance with paragraph (2).

"(2) Any notification required by paragraph (1) with respect to any class or category of vehicles or engines shall be given to dealers, ultimate purchasers, and subsequent purchasers (if known) in such manner and containing such information as the Administrator may by regulations require.

"(3) The manufacturer shall furnish with each new motor vehicle or motor vehicle engine such written instructions for the maintenance and use of the vehicle or engine by the ultimate purchaser as may be reasonable and necessary to assure the proper functioning of emission control devices and systems. In addition, the manufacturer shall indicate by means of a label or tag permanently affixed to such vehicle or engine that such vehicle or engine is covered by a certificate of conformity issued for the purpose of assuring achievement of emissions standards prescribed under section 202. Such label or tag shall contain such other information relating to control of motor vehicle emissions as the Administrator shall prescribe by regulation.

"(d) Any cost obligation of any dealer incurred as a result of any requirement imposed by subsection (a), (b), or (c) shall be borne by the manufacturer. The transfer of any such cost obligation from a manufacturer to any dealer through franchise or other agreement is prohibited.

"(e) If a manufacturer includes in any advertisement a statement respecting the cost or value of emission control devices or systems, such manufacturer shall set forth in such statement the cost or value attributed to such devices or systems by the Secretary of Labor (through the Bureau of Labor Statistics). The Secretary of Labor, and his representatives, shall have the same access for this purpose to the books, documents, papers, and records of a manufacturer as the Comptroller General has to those of a recipient of assistance for purposes of section 311.

"(f) Any inspection of a motor vehicle or a motor vehicle engine for purposes of subsection (c)(1), after its sale to the ultimate purchaser, shall be made only if the owner of such vehicle or engine voluntarily permits such inspection to be made, except as may be provided by any State or local inspection program.

(b) The amendments made by this section shall not apply to vehicles or engines imported into the United States before the sixtieth day after the date of enactment of this Act.

"SEC. 208. (a) Every manufacturer shall establish and maintain such records, make such reports, and provide such information as the Administrator may reasonably require to enable him to determine whether such manufacturer has acted or is acting in compliance with this part and regulations thereunder and shall, upon request of an officer or employee duly designated by the Administrator, permit such officer or employee at reasonable times to have access to and copy such records.

"(b) Any records, reports or information obtained under subsection (a) shall be available to the public, except that upon a showing satisfactory to the Administrator by any person that records, reports, or information, or particular part thereof (other than emission data), to which the Administrator has access under this section if made public, would divulge methods or processes entitled to protection as trade secrets of such person, the Administrator shall consider such record, report, or information or particular portion thereof confidential in accordance with the purposes of section 1905 of title 18 of the United States Code, except that such record, report, or information may be disclosed to other officers, employees, or authorized representatives of the United States concerned with carrying out this Act or when relevant in any proceeding under this Act. Nothing in this section shall authorize the withholding of information by the Administrator or any officer or employee under his control, from the duly authorized committees of the Congress.

"STATE STANDARDS

"SEC. 209. (a) No State or any political subdivision thereof shall adopt or attempt to enforce any standard relating to the control of emissions from new motor vehicles or new motor vehicle engines subject to this part. No State shall require certification, inspection, or any other approval relating to the control of emissions from any new motor vehicle or new motor vehicle engine as condition precedent to the initial retail sale, titling (if any), or registration of such motor vehicle, motor vehicle engine, or equipment.

"(b) The Administrator shall, after notice and opportunity for public hearing, waive application of this section to any State which has adopted standards (other than crankcase emission standards) for the control of emissions from new motor vehicles or new motor vehicle engines prior to March 30, 1966, unless he finds that such State does not require standards more stringent than applicable Federal standards to meet compelling and extraordinary conditions or that such State standards and accompanying enforcement procedures are not consistent with section 202(a) of this part.

"(c) Nothing in this part shall preclude or deny to any State or political subdivision thereof the right otherwise to control, regulate, or restrict the use, operation, or movement of registered or licensed motor vehicles.

"SEC. 210. The Administrator is authorized to make grants to appropriate State agencies in an amount up to two-thirds of the cost of developing and maintaining effective vehicle emission devices and systems inspection and emission testing and control programs, except that—

"(1) no such grant shall be made for any part of any State vehicle inspection program which does not directly relate to the cost of the air pollution control aspects of such a program;

"(2) no such grant shall be made unless the Secretary of Transportation has certified to the Administrator that such program is consistent with any highway safety program developed pursuant to section 402 of title 23 of the United States Code; and

"(3) no such grant shall be made unless the program includes provisions designed to insure that emission control devices and systems on vehicles in actual use have not been discontinued or rendered inoperative.

"REGULATION OF FUELS

"SEC. 211. (a) The Administrator may by regulation designate any fuel or fuel additive and, after such date or dates as may be prescribed by him, no manufacturer or processor of any such fuel or additive may sell, offer for sale, or introduce into commerce such fuel or additive unless the Administrator has registered such fuel or additive in accordance with subsection (b) of this section.

"(b) (1) For the purpose of registration of fuels and fuel additives, the Administrator shall require—

"(A) the manufacturer of any fuel to notify him as to the commercial identifying name and manufacturer of any additive contained in such fuel; the range of concentration of any additive in the fuel; and the purpose-in-use of any such additive; and

"(B) the manufacturer of any additive to notify him as to the chemical composition of such additive.

"(2) For the purpose of registration of fuels and fuel additives, the Administrator may also require the manufacturer of any fuel or fuel additive—

"(A) to conduct tests to determine potential public health effects of such fuel or additive (including, but not limited to, carcinogenic, teratogenic, or mutagenic effects), and

"(B) to furnish the description of any analytical technique that can be used to detect and measure any additive in such fuel, the recommended range of concentration of such additive, and the recommended purpose-in-use of such additive, and such other information as is reasonable and necessary to determine the emissions resulting from the use of the fuel or additive contained in such fuel, the effect of such fuel or additive on the emission control performance of any vehicle or vehicle engine, or the extent to which such emissions affect the public health or welfare.

Tests under subparagraph (A) shall be conducted in conformity with test procedures and protocols established by the Administrator. The result of such tests shall not be considered confidential.

"(3) Upon compliance with the provision of this subsection, including assurances that the Administrator will receive changes in the information required, the Administrator shall register such fuel or fuel additive.

"(c) (1) The Administrator may, from time to time on the basis of information obtained under subsection (b) of this section or other information available to him, by regulation, control or prohibit the manufacture, introduction into commerce, offering for sale, or sale of any fuel or fuel additive for use in a motor vehicle or motor vehicle engine (A) if any emission products of such fuel or fuel additive will endanger the public health or welfare, or (B) if emission products of such fuel or fuel additive will impair to a significant degree the performance of any emission control device or system which is in general use, or which the Administrator finds has been developed to a point where in a reasonable time it would be in general use were such regulation to be promulgated.

"(2) (A) No fuel, class of fuels, or fuel additive may be controlled or prohibited by the Administrator pursuant to clause (A) of paragraph (1) except after consideration of all relevant medical and scientific evidence available to him, including consideration of other technologically or economically feasible means of achieving emission standards under section 202.

"(B) No fuel or fuel additive may be controlled or prohibited by the Administrator pursuant to clause (B) of paragraph (1) except after consideration of available scientific and economic data, including a cost benefit analysis comparing emission control devices or systems which are or will be in general use and require the proposed control or prohibition with emission control devices or systems which are or will be in general use and do not require the proposed control or prohibition. On request of a manufacturer of motor vehicles, motor vehicle engines, fuels, or fuel additives submitted within 10 days of notice of proposed rulemaking, the Administrator shall hold a public hearing and publish findings with respect to any matter he is required to consider under this subparagraph. Such findings shall be published at the time of promulgation of final regulations.

"(C) No fuel or fuel additive may be prohibited by the Administrator under paragraph (1) unless he finds, and publishes such finding, that in his judgment such prohibition will not cause the use of any other fuel or fuel additive which will produce emissions which will endanger the public health or welfare to the same or greater degree than the use of the fuel or fuel additive proposed to be prohibited.

"(3) (A) For the purpose of evidence and data to carry out paragraph (2), the Administrator may require the manufacturer of any motor vehicle or motor vehicle engine to furnish any information which has been developed concerning the emissions from motor vehicles resulting from the use of any fuel or fuel additive, or the effect of such use on the performance of any emission control device or system.

"(B) In obtaining information under subparagraph (A), section 307(a) (relating to subpenas) shall be applicable.

"(4)(A) Except as otherwise provided in subparagraph (B) or (C), no State (or political subdivision thereof) may prescribe or attempt to enforce, for purposes of motor vehicle emission control, any control or prohibition respecting use of a fuel or fuel additive in a motor vehicle or motor vehicle engine—

"(i) if the Administrator has found that no control or prohibition under paragraph (1) is necessary and has published his finding in the Federal Register, or

"(ii) if the Administrator has prescribed under paragraph (1) a control or prohibition applicable to such fuel or fuel additive, unless State prohibition or control is identical to the prohibition or control prescribed by the Administrator.

"(B) Any State for which application of section 209(a) has at any time been waived under section 209(b) may at any time prescribe and enforce, for the purpose of motor vehicle emission control, a control or prohibition respecting any fuel or fuel additive.

"(C) A State may prescribe and enforce, for purposes of motor vehicle emission control, a control or prohibition respecting the use of a fuel or fuel additive in a motor vehicle or motor vehicle engine if an applicable implementation plan for such State under section 110 so provides. The Administrator may approve such provision in an implementation plan, or promulgate an implementation plan containing such a provision, only if he finds that the State control or prohibition is necessary to achieve the national primary or secondary ambient air quality standard which the plan implements.

"(d) Any person who violates subsection (a) or the regulations prescribed under subsection (c) or who fails to furnish any information required by the Administrator under subsection (c) shall forfeit and pay to the United States a civil penalty of $10,000 for each and every day of the continuance of such violation, which shall accrue to the United States and be recovered in a civil suit in the name of the United States, brought in the district where such person has his principal office or in any district in which he does business. The Administrator may, upon application therefor, remit or mitigate any forfeiture provided for in this subsection and he shall have authority to determine the facts upon all such applications.

"DEVELOPMENT OF LOW-EMISSION VEHICLES

"SEC. 212. (a) For the purpose of this section—

"(1) The term 'Board' means the Low-Emission Vehicle Certification Board.

"(2) The term 'Federal Government' includes the legislative, executive, and judicial branches of the Government of the United States, and the government of the District of Columbia.

"(3) The term 'motor vehicle' means any self-propelled vehicle designed for use in the United States on the highways, other than a vehicle designed or used for military field training, combat, or tactical purposes.

"(4) The term 'low-emission vehicle' means any motor vehicle which—

"(A) emits any air pollutant in amounts significantly below new motor vehicle standards applicable under section 202 at the time of procurement to that type of vehicle; and

"(B) with respect to all other air pollutants meets the new motor vehicle standards applicable under section 202 at the time of procurement to that type of vehicle.

"(5) The term 'retail price' means (A) the maximum statutory price applicable to any class or model of motor vehicle; or (B) in any case where there is no applicable maximum statutory price, the most recent procurement price paid for any class or model of motor vehicle.

"(b)(1) There is established a Low-Emission Vehicle Certification Board to be composed of the Administrator or his designee, the Secretary of Transportation or his designee, the Chairman of the Council on Environmental Quality or his designee, the Director of the National Highway Safety Bureau in the Department of Transportation, the Administrator of General Services, and two members appointed by the President. The President shall designate one member of the Board as Chairman.

"(2) Any member of the Board not employed by the United States may receive compensation at the rate of $125 for each day such member is engaged upon work of the Board. Each member of the Board shall be reimbursed for travel expenses, including per diem in lieu of subsistence as authorized by section 5703 of title 5, United States Code, for persons in the Government service employed intermittently.

"(3)(A) The Chairman, with the concurrence of the members of the Board, may employ and fix the compensation of such additional personnel as may be necessary to carry out the functions of the Board, but no individual so appointed shall receive compensation in excess of the rate authorized for GS–18 by section 5332 of title 5, United States Code.

"(B) The Chairman may fix the time and place of such meetings as may be required, but a meeting of the Board shall be called whenever a majority of its members so request.

"(C) The Board is granted all other powers necessary for meeting its responsibilities under this section.

"(c) The Administrator shall determine which models or classes of motor vehicles qualify as low-emission vehicles in accordance with the provisions of this section.

"(d)(1) The Board shall certify any class or model of motor vehicles—

"(j) The Board shall promulgate the procedures required to implement this section within one hundred and eighty days after the date of enactment of the Clean Air Amendments of 1970.

"DEFINITIONS FOR PART A

"SEC. 213. As used in this part—

"(1) The term 'manufacturer' as used in sections 202, 203, 206, 207, and 208 means any person engaged in the manufacturing or assembling of new motor vehicles or new motor vehicle engines, or importing such vehicles or engines for resale, or who acts for and is under the control of any such person in connection with the distribution of new motor vehicles or new motor vehicle engines, but shall not include any dealer with respect to new motor vehicles or new motor vehicle engines received by him in commerce.

"(2) The term 'motor vehicle' means any self-propelled vehicle designed for transporting persons or property on a street or highway.

"(3) Except with respect to vehicles or engines imported or offered for importation, the term 'new motor vehicle' means a motor vehicle the equitable or legal title to which has never been transferred to an ultimate purchaser; and the term 'new motor vehicle engine' means an engine in a new motor vehicle or a motor vehicle engine the equitable or legal title to which has never been transferred to the ultimate purchaser; and with respect to imported vehicles or engines, such terms mean a motor vehicle and engine, respectively, manufactured after the effective date of a regulation issued under section 202 which is applicable to such vehicle or engine (or which would be applicable to such vehicle or engine had it been manufactured for importation into the United States).

"(4) The term 'dealer' means any person who is engaged in the sale or the distribution of new motor vehicles or new motor vehicle engines to the ultimate purchaser.

"(5) The term 'ultimate purchaser' means, with respect to any new motor vehicle or new motor vehicle engine, the first person who in good faith purchases such new motor vehicle or new engine for purposes other than resale.

"(6) The term 'commerce' means (A) commerce between any place in any State and any place outside thereof; and (B) commerce wholly within the District of Columbia.

"PART B—AIRCRAFT EMISSION STANDARDS

"ESTABLISHMENT OF STANDARDS

"SEC. 231 (a) (1) Within 90 days after the date of enactment of the Clean Air Amendments of 1970, the Administrator shall commence a study and investigation of emissions of air pollutants from aircraft in order to determine—

"(A) the extent to which such emissions affect air quality in air quality control regions throughout the United States, and

"(B) the technological feasibility of controlling such emissions.

"(2) Within 180 days after commencing such study and investigation, the Administrator shall publish a report of such study and investigation and shall issue proposed emission standards applicable to emissions of any air pollutant from any class or classes of aircraft or aircraft engines which in his judgment cause or contribute to or are likely to cause or contribute to air pollution which endangers the public health or welfare.

"(3) The Administrator shall hold public hearings with respect to such proposer standards. Such hearings shall, to the extent practicable, be held in air quality control regions which are most seriously affected by aircraft emissions. Within 90 days after the issuance of such proposed regulations, he shall issue such regulation with such modifications as he deems appropriate. Such regulations may be revised from time to time.

"(b) Any regulation prescribed under this section (and any revision thereof) shall take effect after such period as the Administrator finds necessary (after consultation with the Secretary of Transportation) to permit the development and application of the requisite technology, giving appropriate consideration to the cost of compliance within such period.

"(c) Any regulations under this section, or amendments thereto, with respect to aircraft, shall be prescribed only after consultation with the Secretary of Transportation in order to assure appropriate consideration for aircraft safety.

"ENFORCEMENT OF STANDARDS

"SEC. 232 (a) The Secretary of Transportation, after consultation with the Administrator, shall prescribe regulations to insure compliance with all standards prescribed under section 231 by the Administrator. The regulations of the Secretary of Transportation shall include provisions making such standards applicable in the issuance, amendment, modification, suspension, or revocation of any certificate authorized by the Federal Aviation Act or the Department of Transportation Act. Such Secretary shall insure that all necessary inspections are accomplished, and, may execute any power or duty vested in him by any other provision of law in the execution of all powers and duties vested in him under this section.

"(b) In any action to amend, modify, suspend, or revoke a certificate in which violation of an emission standard prescribed under section 231 or of a regulation prescribed under subsection (a) is at issue, the certificate holder shall have the same notice and appeal rights as are prescribed for such holders in the Federal Aviation Act of 1958 or the Department of Transportation Act, except that in any appeal to the National Transportation Safety Board, the Board may amend, modify, or revoke the order of the

Secretary of Transportation only if it finds no violation of such standard or regulation and that such amendment, modification, or revocation is consistent with safety in air transportation.

"STATE STANDARDS AND CONTROLS

"SEC. 233. No State or political subdivision thereof may adopt or attempt to enforce any standard respecting emissions of any air pollutant from any aircraft or engine thereof unless such standard is identical to a standard applicable to such aircraft under this part.

"DEFINITIONS

"SEC. 234. Terms used in this part (other than Administrator) shall have the same meaning as such terms have under section 101 of the Federal Aviation Act of 1958.

"TITLE III—GENERAL

"ADMINISTRATION

"SEC. 301. (a) The Administrator is authorized to prescribe such regulations as are necessary to carry out his functions under this Act. The Administrator may delegate to any officer or employee of the Environmental Protection Agency such of his powers and duties under this Act, except the making of regulations, as he may deem necessary or expedient.

"(b) Upon the request of an air pollution control agency, personnel of the Environmental Protection Agency may be detailed to such agency for the purpose of carrying out the provisions of this Act.

"(c) Payments under grants made under this Act may be made in installments, and in advance or by way of reimbursement, as may be determined by the Administrator.

"DEFINITIONS

"SEC. 302. When used in this Act—

"(a) The term 'Administrator' means the Administrator of the Environmental Protection Agency.

"(b) The term 'air pollution control agency' means any of the following:

"(1) A single State agency designated by the Governor of that State as the official State air pollution control agency for purposes of this Act;

"(2) An agency established by two or more States and having substantial powers or duties pertaining to the prevention and control of air pollution;

"(3) A city, county, or other local government health authority, or, in the case of any city, county, or other local government in which there is an agency other than the health authority charged with responsibility for enforcing ordinances or laws relating to the prevention and control of air pollution, such other agency; or

"(4) An agency of two or more municipalities located in the same State or in different States and having substantial powers or duties pertaining to the prevention and control of air pollution.

"(c) The term 'interstate air pollution control agency' means—

"(1) an air pollution control agency established by two or more States, or

"(2) an air pollution control agency of two or more municipalities located in different States.

"(d) The term 'State' means a State, the District of Columbia, the Commonwealth of Puerto Rico, the Virgin Islands, Guam, and American Samoa.

"(e) The term 'person' includes an individual, corporation, partnership, association, State, municipality, and political subdivision of a State.

"(f) The term 'municipality' means a city, town, borough, county, parish, district, or other public body created by or pursuant to State law.

"(g) The term 'air pollutant' means an air pollution agent or combination of such agents.

"(h) All language referring to effects on welfare includes, but is not limited to, effects on soils, water, crops, vegetation, manmade materials, animals, wildlife, weather, visibility, and climate, damage to and deterioration of property, and hazards to transportation, as well as effects on economic values and on personal comfort and well-being.

"EMERGENCY POWERS

"SEC. 303. Notwithstanding any other provisions of this Act, the Administrator upon receipt of evidence that a pollution source or combination of sources (including moving sources) is presenting an imminent and substantial endangerment to the health of persons, and that appropriate State or local authorities have not acted to abate such sources, may bring suit on behalf of the United States in the appropriate United States district court to immediately restrain any person causing or contributing to the alleged pollution to stop the emission of air pollutants causing or contributing to such pollution or to take such other action as may be necessary.

"CITIZEN SUITS

"SEC. 304. (a) Except as provided in subsection (b), any person may commence a civil action on his own behalf—

"(1) against any person (including (i) the United States, and (ii) any other governmental instrumentality or agency to the extent permitted by the Eleventh Amendment to the

68

Constitution) who is alleged to be in violation of (A) an emission standard or limitation under this Act or (B) an order issued by the Administrator or a State with respect to such a standard or limitation, or

"(2) against the Administrator where there is alleged a failure of the Administrator to perform any act or duty under this Act which is not discretionary with the Administrator.

The district courts shall have jurisdiction, without regard to the amount in controversy or the citizenship of the parties, to enforce such an emission standard or limitation, or such an order, or to order the Administrator to perform such act or duty, as the case may be.

"(b) No action may be commenced—

"(1) under subsection (a) (1)—

"(A) prior to 60 days after the plaintiff has given notice of the violation (i) to the Administrator, (ii) to the State in which the violation occurs, and (iii) to any alleged violator of the standard, limitation, or order, or

"(B) if the Administrator or State has commenced and is diligently prosecuting a civil action in a court of the United States or a State to require compliance with the standard, limitation, or order, but in any such action in a court of the United States any person may intervene as a matter of right.

"(2) under subsection (a) (2) prior to 60 days after the plaintiff has given notice of such action to the Administrator, except that such action may be brought immediately after such notification in the case of an action under this section respecting a violation of section 112(c) (1) (B) or an order issued by the Administrator pursuant to section 113(a). Notice under this subsection shall be given in such manner as the Administrator shall prescribe by regulation.

"(c) (1) Any action respecting a violation by a stationary source of an emission standard or limitation or an order respecting such standard or limitation may be brought only in the judicial district in which such source is located.

"(2) In such action under this section, the Administrator, if not a party, may intervene as a matter of right.

"(d) The court, in issuing any final order in any action brought pursuant to subsection (a) of this section, may award costs of litigation (including reasonable attorney and expert witness fees) to any party, whenever the court determines such award is appropriate. The court may, if a temporary restraining order or preliminary injunction is sought, require the filing of a bond or equivalent security in accordance with the Federal Rules of Civil Procedure.

"(e) Nothing in this section shall restrict any right which any person (or class of persons) may have under any statute or common law to seek enforcement of any emission standard or limitation or to seek any other relief (including relief against the Administrator or a State agency).

"(f) For purposes of this section, the term 'emission standard or limitation under this Act' means—

"(1) a schedule or timetable of compliance, emission limitation, standard of performance or emission standard, or

"(2) a control or prohibition respecting a motor vehicle fuel or fuel additive,

which is in effect under this Act (including a requirement applicable by reason of section 118) or under an applicable implementation plan.

"APPEARANCE

"SEC. 305. The Administrator shall request the Attorney General to appear and represent him in any civil action instituted under this Act to which the Administrator is a party. Unless the Attorney General notifies the Administrator that he will appear in such action, within a reasonable time, attorneys appointed by the Administrator shall appear and represent him.

"FEDERAL PROCUREMENT

"SEC. 306. (a) No Federal agency may enter into any contract with any person who is convicted of any offense under section 113(c)(1) for the procurement of goods, materials, and services to perform such contract at any facility at which the violation which gave rise to such conviction occurred if such facility is owned, leased, or supervised by such person. The prohibition in the preceding sentence shall continue until the Administrator certifies that the condition giving rise to such a conviction has been corrected.

"(b) The Administrator shall establish procedures to provide all Federal agencies with the notification necessary for the purposes of subsection (a).

"(c) In order to implement the purposes and policy of this Act to protect and enhance the quality of the Nation's air, the President shall, not more than 180 days after enactment of the Clean Air Amendments of 1970 cause to be issued an order (1) requiring each Federal agency authorized to enter into contracts and each Federal agency which is empowered to extend Federal assistance by way of grant, loan, or contract to effectuate the purpose and policy of this Act in such contracting or assistance activities, and (2) setting forth procedures, sanctions, penalties, and such other provisions, as the President determines necessary to carry out such requirement.

"(d) The President may exempt any contract, loan, or grant from all or part of the provisions of this section where he determines such exemption is necessary in the paramount interest of the United States and he shall notify the Congress of such exemption.

"(e) The President shall annually report to the Congress on measures taken toward implementing the purpose and intent of this section, including but not limited to the progress and problems associated with implementation of this section.

"GENERAL PROVISION RELATING TO ADMINSITRATIVE PROCEEDINGS AND JUDICIAL REVIEW

"SEC. 307 (a) (1) In connection with any determination under section 110(f) or section 202(b) (5), or for purposes of obtaining information under section 202(b) (4) or 210(c) (4), the Administrator may issue subpenas for the attendance and testimony of witnesses and the production of relevant papers, books, and documents, and he may administer oaths. Except for emission data, upon a showing satisfactory to the Administrator by such owner or operator that such papers, books, documents, or information or particular part thereof, if made public, would divulge trade secrets or secret processes of such owner or operator, the Administrator shall consider such record, report, or information or particular portion thereof confidential in accordance with the purposes of section 1905 of title 18 of the United States Code, except that such paper, book, document, or information may be disclosed to other officers, emplyees, or authorized representatives of the United States concerned with carrying out this Act, to persons carrying out the National Academy of Sciences' study and investigation provided for in section 202(c), or when relevant in any proceeding under this Act. Witnesses summoned shall be paid the same fees and mileage that are paid witnesses in the courts of the United States. In cases of contumacy or refusal to obey a supena served upon any person under this subparagraph, the district court of the United States for any district in which such person is found or resides or transacts business, upon application by the United States and after notice to such person, shall have jurisdiction to issue an order requiring such person to appear and give testimony before the Administrator to appear and produce papers, books, and documents before the Administrator, or both, and any failure to obey such order of the court may be punished by such court as a contempt thereof.

"(b) (1) A petition for review of action of the Administrator in promulgating any national primary or secondary ambient air quality standard, any emission standard under section 112, any standard of performance under section 111 any standard under section 202 (other than a standard required to be prescribed under section 202(b) (1)), any determination under section 202(b) (5), any control or prohibition under section 211, or any standard under section 231 may be filed only in the United States Court of Appeals for the District of Columbia. A petition for review of the Administrator's action in approving or promulgating any implementation plan under section 110 or section 111(d) may be filed only in the United States Court of Appeals for the appropriate circuit. Any such petition shall be filed within 30 days from the date of such promulgation or approval, or after such date if such petition is based solely on grounds arising after such 30th day.

"(2) Action of the Administrator with respect to which review could have been obtained under paragraph (1) shall not be subject to judicial review in civil or criminal proceedings for enforcement.

"(c) In any judicial proceeding in which review is sought of a determination under this Act required to be made on the record after notice and opportunity for hearing, if any party applies to the court for leave to adduce additional evidence, and shows to the satisfaction of the court that such additional evidence is material and that there were reasonable grounds for the failure to adduce such evidence in the proceeding before the Administrator, the court may order such additional evidence (and evidence in rebuttal thereof) to be taken before the Administrator, in such manner and upon such terms and conditions as to the court may deem proper. The Administrator may modify his findings as to the facts, or make new findings, by reason of the additional evidence so taken and he shall file such modified or new findings, and his recommendation, if any, for the modification or setting aside of his original determination, with the return of such additional evidence.

"MANDATORY LICENSING

"SEC. 308. Whenever the Attorney General determines, upon application of the Administrator—
    "(1) that—
        "(A) in the implementation of the requirements of section 111, 112, or 202 of this Act, a right under any United States letters patent, which is being used or intended for public or commercial use and not otherwise reasonably available, is necessary to enable any person required to comply with such limitation to so comply, and
        "(B) there are no reasonable alternative methods to accomplish such purpose, and
    "(2) that the unavailability of such right may result in a substantial lessening of competition or tendency to create a monopoly in any line of commerce in any section of the country,
the Attorney General may so certify to a district court of the United States, which may issue an order requiring the person who owns such patent to license it on such reasonable terms and conditions as the court, after hearing, may determine. Such certification may be made to the district court for the district in which the person owning the patent resides, does business, or is found.

"POLICY REVIEW

"SEC. 309. (a) The Administrator shall review and comment in writing on the environmental impact of any matter relating to duties and responsibilities granted pursuant to this Act or other provisions of the authority of the Administrator, contained in any (1) legislation proposed by any Federal department or agency, (2) newly authorized Federal projects for construction and any major Federal agency action other than a project for construction to which section 102(2)(C) of Public Law 91–190 applies, and (3) proposed regulations published by any department or agency of the Federal Government. Such written comment shall be made public at the conclusion of any such review.

"(b) In the event the Administrator determines that any such legislation, action, or regulation is unsatisfactory from the standpoint of public health or welfare or environmental quality, he shall publish his determination and the matter shall be referred to the Council on Environmental Quality.

"OTHER AUTHORITY NOT AFFECTED

"SEC. 310. (a) Except as provided in subsection (b) of this section, this Act shall not be construed as superseding or limiting the authorities and responsibilities, under any other provision of law, of the Administrator or any other Federal officer, department, or agency.

"(b) No appropriation shall be authorized or made under section 301, 311, or 314 of the Public Health Service Act for any fiscal year after the fiscal year ending June 30, 1964, for any purpose for which appropriations may be made under authority of this Act.

"RECORDS AND AUDIT

"SEC. 311. (a) Each recipient of assistance under this Act shall keep such records as the Administrator shall prescribe, including records which fully disclose the amount and disposition by such recipient of the proceeds of such assistance, the total cost of the project or undertaking in connection with which such assistance is given or used, and the amount of that portion of the cost of the project or undertaking supplied by other sources, and such other records as will facilitate an effective audit.

"(b) The Administrator and the Comptroller General of the United States, or any of their duly authorized representatives, shall have access for the purpose of audit and examinations to any books, documents, papers, and records of the recipients that are pertinent to the grants received under this Act.

"COMPREHENSIVE ECONOMIC COST STUDIES

"SEC. 312. (a) In order to provide the basis for evaluating programs authorized by this Act and the development of new programs and to furnish the Congress with the information necessary for authorization of appropriations by fiscal years beginning after June 30, 1969, the Administrator, in cooperation with State, interstate, and local air pollution control agencies, shall make a detailed estimate of the cost of carrying out the provisions of this Act; a comprehensive study of the cost of program implementation by affected units of government; and a comprehensive study of the economic impact of air quality standards on the Nation's industries, communities, and other contributing sources of pollution, including an analysis of the national requirements for and the cost of controlling emissions to attain such standards of air quality as may be established pursuant to this Act or applicable State law. The Administrator shall submit such detailed estimate and the results of such comprehensive study of cost for the five-year period beginning July 1, 1969, and the results of such other studies, to the Congress not later than January 10, 1969, and shall submit a reevaluation of such estimate and studies annually thereafter.

"(b) The Administrator shall also make a complete investigation and study to determine (1) the need for additional trained State and local personnel to carry out programs assisted pursuant to this Act and other programs for the same purpose as this Act; (2) means of using existing Federal training programs to train such personnel; and (3) the need for additional trained personnel to develop, operate and maintain those pollution control facilities designed and installed to implement air quality standards. He shall report the results of such investigation and study to the President and the Congress not later than July 1, 1969.

"ADDITIONAL REPORTS TO CONGRESS

"SEC. 313. Not later than six months after the effective date of this section and not later than January 10 of each calendar year beginning after such date, the Administrator shall report to the Congress on measures taken toward implementing the purpose and intent of this Act including, but not limited to, (1) the progress and problems associated with control of automotive exhaust emissions and the research efforts related thereto; (2) the development of air quality criteria and recommended emission control requirements; (3) the status of enforcement actions taken pursuant to this Act; (4) the status of State ambient air standards setting, including such plans for implementation and enforcement as have been developed; (5) the extent of development and expansion of air pollution monitoring systems; (6) progress and problems related to development of new and improved control techniques; (7) the develpment of quantitative and qualitative instrumentation to monitor emissions and air quality; (8) standards set or under consideration pursuant to title II of this Act; (9) the status of State, interstate, and local pollution control programs established pursuant to and assisted by this Act; and (10) the reports and recommendations made by the President's Air Quality Advisory Board.

"LABOR STANDARDS

"SEC. 314. The Administrator shall take such action as may be necessary to insure that all laborers and mechanics employed by contractors or subcontractors on projects assisted under this Act shall be paid wages at rates not less than those prevailing for the same type of work on similar construction in the locality as determined by the Secretary of Labor, in accordance with the Act of March 3, 1931, as amended, known as the Davis-Bacon Act (46 Stat. 1494; 40 U.S.C. 276a—276a-5). The Secretary of Labor shall have, with respect to the labor standards specified in this subsection, the authority and functions set forth in Reorganization Plan Numbered 14 of 1950 (15 F.R. 3176; 64 Stat. 1267) and section 2 of the Act of June 13, 1934, as amended (48 Stat. 948; 40 U.S.C. 276c).

"SEPARABILITY

"SEC. 315. If any provision of this Act, or the application of any provision of this Act to any person or circumstance, is held invalid, the application of such provision to other persons or circumstances, and the remainder of this Act, shall not be affected thereby.

"APPROPRIATIONS

"SEC. 316. There are authorized to be appropriated to carry out this Act, other than sections 103(f)(3) and (d), 104, 212, and 403, $125,000,000 for the fiscal year ending June 30, 1971, $225,000,000 for the fiscal year ending June 30, 1972, and $300,000,000 for the fiscal year ending June 30, 1973.

SAVINGS PROVISIONS [1]

SEC. 16. (a) (1) Any implementation plan adopted by any State and submitted to the Secretary of Health, Education, and Welfare, or to the Administrator pursuant to the Clean Air Act prior to enactment of this Act may be approved under section 110 of the Clean Air Act (as amended by this Act) and shall remain in effect, unless the Administrator determines that such implementation plan, or any portion thereof, is not consistent with the applicable requirements of the Clean Air Act (as amended by this Act) and will not provide for the attainment of national primary ambient air quality standards in the time required by such Act. If the Administrator so determines, he shall, within ninety days after promulgation of any national ambient air quality standards pursuant to section 109(a) of the Clean Air Act, notify the State and specify in what respects changes are needed to meet the additional requirements of such Act, including requirements to implement national secondary ambient air quality standards. If such changes are not adopted by the State after public hearings and within six months after such notification, the Administrator shall promulgate such changes pursuant to section 110(c) of such Act.

(2) The amendments made by section 4(b) shall not be construed as repealing or modifying the powers of the Administrator

---

[1] Provisions included in Clean Air Amendments of 1970. In these provisions, the phrases "prior to enactment of this Act" and "as amended by this Act" refer to enactment of the Clean Air Amendments of 1970.

with respect to any conference convened under section 108(d) of the Clean Air Act before the date of enactment of this Act.[2]

(b) Regulations or standards issued under title II of the Clean Air Act prior to the enactment of this Act shall continue in effect until revised by the Administrator consistent with the purposes of such Act.

(1) Section 601 of the Federal Aviation Act of 1958 (49 U.S.C. 1421) is amended by adding at the end thereof the following new subsection:

"AVIATION FUEL STANDARDS [1]

"(d) The Administrator shall prescribe, and from time to time revise, regulations (1) establishing standards governing the composition or the chemical or physical properties of any aircraft fuel or fuel additive for the purpose of controlling or eliminating aircraft emissions which the Administrator of the Environmental Protection Agency (pursuant to section 231 of the Clean Air Act) determines endanger the public health or welfare, and (2) providing for the implementation and enforcement of such standards."

(2) Section 610(a) of such Act (49 U.S.C. 1430(a)) is amended by striking out "and" at the end of paragraph (7); by striking out the period at the end of paragraph (8) and inserting in lieu thereof "; and" and by adding after paragraph (8) the following new paragraph:

"(9) For any person to manufacture, deliver, sell, or offer for sale, any aviation fuel or fuel additive in violation of any regulation prescribed under section 601(d)."

(3) That portion of the table of contents contained in the first section of the Federal Aviation Act of 1958 which appears under the side heading

"SEC. 601 General Safety Powers and Duties."

is amended by adding at the end thereof the following:

"(d) Aviation fuel standards."

[2] The amendments referred to in this paragraph were contained in section 4(b) of the Clean Air Amendments of 1970. They are reflected in the provisions of what is now section 115 of the Clean Air Act.

[1] These amendments to the Federal Aviation Act were made by the Clean Air Amendments of 1970 and are included herein because of their relationship to the Clean Air Act.

Summary

The Clean Air Act was enacted to protect and improve the quality of the nation's air resources so as to promote the public health, welfare, and productive capacity of its population. In addition, the Act has served to initiate and accelerate a national research and development program to achieve the prevention and control of air pollution. Technical and financial assistance to State and local governments in conjunction with the development and execution of their air prevention and control programs, as well as encouragement and assistance in the development and operation of regional air pollution control programs, is also provided under the Act.

Review Questions

1. What are the basic purposes of the Clean Air Act?

2. Identify one economic factor that aided the passage of the Clean Air Act?

3. Identify one political factor that aided the passage of the Clean Air Act?

4. Identify one social factor that aided the passage of the Clean Air Act?

5. What effect has the Clean Air Act had on the automotive industry since 1963?

6. What are two strengths of the Clean Air Act with regard to current pollution control problems?

7. What are two weaknesses of the Clean Air Act with regard to current pollution control problems?

8. What are two strengths of the Clean Air Act with regard to current energy conservation measures?

9. What are two weaknesses of the Clean Air Act with regard to current energy conservation measures?

solid waste management

OBJECTIVES:

The student should be able to identify the specific purposes of the Solid Waste Disposal Act.

The student should be able to describe those economic, political, and social factors leading to the passage of the Solid Waste Disposal Act.

The student should be able to describe the effect the Solid Waste Disposal Act has had on various types of industries.

The student should be able to identify the strengths and weaknesses of the Solid Waste Disposal Act as they relate to current pollution control problems and energy conservation measures.

INTRODUCTION:

While accounting for only 7 percent of the world's population, Americans consume nearly half of the earth's industrial raw materials. Not surprisingly, the way of life that requires such large amounts of natural resources also produces enormous amounts of wastes in the solid state. But until recently, Americans were not greatly concerned with environmental problems associated with the collection and disposal of trash, garbage, or other solid wastes. In a vast country, with low population density and seemingly unlimited natural resources, the most convenient disposal method—usually an open dump—seemed adequate. There appeared no reason to reuse wastes since virgin materials were abundant and often cheaper than reclaimed materials.

TOWARD A NEW ENVIRONMENTAL ETHIC, U.S. Government Printing Office: 1971-O — 443-062, pp. 18-19.

Today, however, this view has been replaced by a genuine concern, not only for improved disposal methods but for the recovery and reuse of the valuable and often irreplaceable resources that form a large part of the discards of this high-production, high-consumption society.

The solid wastes produced in the United States now total 4.3 billion tons a year. Of this, 360 million tons are household, municipal, and industrial wastes. In addition there are 2.3 billion tons of agricultural wastes and 1.7 billion tons of mineral wastes.

Of this annual total 190 million tons, or 5.3 pounds per person per day, are picked up by some collection agency and hauled away for disposal —at a cost of over $4.5 billion per year.

Most present disposal methods pollute either land, air, or water. Three-fourths of the dumps contribute to air pollution and half of them are so situated that their drainage aggravates pollution of rivers and streams. Almost all municipal incinerators are obsolescent in terms of today's needs and technology.

A national survey has revealed that less than 6 percent of 12,000 land disposal sites meet the minimum federal standards for sanitary landfills; and all over the country, cities, unable to find convenient space for land disposal, are desperately seeking new sites—even distant sites—to which they can haul trainloads of municipal wastes. In the cities, all too frequently, inadequate collection results in waste accumulations that breed disease, rats and accidents.

By 1980, it is expected that waste collection will mount to over 340 millions tons per year, or 8 pounds per person per day. It is estimated, in fact, that our solid waste load is presently increasing at twice the rate of the population increase.

Americans see the effect of the present solid waste load everywhere: in smoking dumps that ring their cities and add to the pollution of their air; in the foul-smelling barges that make their way out to sea to dump their cargoes of sludge; in the overflowing garbage cans that line their sidewalks or alleys; in the acres of junked car graveyards that mar their countryside.

79

The annual "throwaway" includes 48 billion cans, 26 billion bottles and jars, 4 million tons of plastic, 7.6 million television sets, 7 million cars and trucks and 30 million tons of paper. The problems of disposal have been aggravated by widespread and increasing use of packaging, disposable containers, and other convenience materials that do not burn or decay.

But the environmental pollution, the scenic blight, the waste disposal difficulties are only part of the total problem. The vast quantities of non-renewable resources, such as ferrous metals, which are permanently lost in the solid waste stream, present a growing and unnecessary economic and resource drain.

Today, a new concept of solid waste management is evolving; it assumes that man can devise a social-technological system that will wisely control the quantity and characteristics of wastes, efficiently collect those that must be removed, creatively recycle those that can be reused, and properly dispose of those that have no further use.

The Solid Waste Disposal Act of 1965 marked the first significant interest by the federal government in management of solid wastes. The act provided for assistance to state and local governments, and others involved in managing solid wastes, by financial grants to demonstrate new technology, technical assistance through research and training and by encouragement of proper planning for state and local solid waste management programs.

The Resource Recovery Act of 1970 amended the legislation to provide a new focus on recycling and recovery of valuable waste materials. Under current legislation, EPA:

• Performs research to find improved methods in all aspects of solid waste management and provides technical assistance to speed the application of new knowledge.

Special emphasis is given to studies to determine means of recovering materials and energy from solid waste; methods of accelerating the reclamation of such materials (by economic incentives and disincentives, subsidies, depletion allowances, federal procurement to develop market demand, etc.); and the feasibility of reducing

80

the amount of solid wastes by changes in product characteristics, production or packaging practices.

• Makes financial grants for the construction and operation of plants or processes for demonstrating new technology.

In the city of Franklin, Ohio, for example, an advanced system for recovery of municipal wastes is being demonstrated. It features a hydropulper, by which solid wastes are processed into slurry form. Heavy materials are ejected and ferrous metals removed for salvage by an electromagnet. Paper fiber is recovered for reuse. An additional step planned involves extraction of glass, with separation into various colors by an optical sorting device. The residue has a relatively high percentage of aluminum, which also may be reclaimed.

• Is developing a comprehensive plan for a system of national disposal sites for storage and disposal of hazardous wastes.

• Provides financial assistance to state and local governments and interstate agencies for the development of resource recovery and solid waste disposal systems and for solid waste management planning. By 1971, 50 state or interstate agencies have used this assistance for developing statewide or regional plans for managing solid wastes.

• Provides training to develop the highly skilled engineers and technicians needed to design, operate, and maintain complex new regional systems.

SOLID WASTE DISPOSAL ACT

[PUBLIC LAW 89–272—89TH CONGRESS, S. 306, APPROVED
OCTOBER 20, 1965]

AN ACT To authorize a research and development program with
respect to solid-waste disposal, and for other purposes.

*　　　*　　　*　　　*　　　*

TITLE II—SOLID WASTE DISPOSAL [1]

SHORT TITLE

SEC. 201. This title (hereinafter referred to as "this Act") may be cited as the "Solid Waste Disposal Act"

Solid Waste
Disposal Act

FINDINGS AND PURPOSES

SEC. 202. (a) The Congress finds—

(1) that the continuing technological progress and improvement in methods of manufacture, packaging, and marketing of consumer products has resulted in an ever-mounting increase, and in a change in the characteristics, of the mass of material discarded by the purchaser of such products;

(2) that the economic and population growth of our Nation, and the improvements in the standard of living enjoyed by our population, have required increased industrial production to meet our needs, and have made necessary the demolition of old buildings, the construction of new buildings, and the provision of highways and other avenues of transportation, which, together with related industrial, commercial, and agricultural operations, have resulted in a rising tide of scrap, discarded, and waste materials;

(3) that the continuing concentration of our population in expanding metropolitan and other urban areas has presented these communities with serious financial, management, intergovernmental, and technical problems in the disposal of solid wastes resulting from the industrial, commercial, domestic, and other activities carried on in such areas;

(4) that inefficient and improper methods of disposal of solid wastes result in scenic blights, create serious hazards to the public health, including pollution of air and water resources, accident hazards, and increase in rodent and insect vectors of disease,

---

[1] Title I of P.L. 89–272 amended the Clean Air Act (P.L. 88–206).

THE SOLID WASTE DISPOSAL ACT, 3rd Revision, 1973, U.S. Government Printing Office: 1973 — 759-907/1131.

have an adverse effect on land values, create public nuisances, otherwise interfere with community life and development;

(5) that the failure or inability to salvage and re-use such materials economically results in the unnecessary waste and depletion of our natural resources; and

(6) that while the collection and disposal of solid wastes should continue to be primarily the function of State, regional, and local agencies, the problems of waste disposal as set forth above have become a matter national in scope and in concern and necessitate Federal action through financial and technical assistance and leadership in the development, demonstration, and application of new and improved methods and processes to reduce the amount of waste and unsalvageable materials and to provide for proper and economical solid-waste disposal practices.

(b)[2] The purposes of this Act therefore are—

(1) to promote the demonstration, construction, and application of solid waste management and resource recovery systems which preserve and enhance the quality or air, water, and land resources;

(2) to provide technical and financial assistance to States and local governments and interstate agencies in the planning and development of resource recovery and solid waste disposal programs;

(3) to promote a national research and development program for improved management techniques, more effective organizational arrangements, and new and improved methods of collection, separation, recovery, and recycling of solid wastes, and the environmentally safe disposal of nonrecoverable residues;

(4) to provide for the promulgation of guidelines for solid waste collection, transport, separation, recovery, and disposal systems; and

(5) to provide for training grants in occupations involving the design, operation, and maintenance of solid waste disposal systems.

DEFINITIONS

Sec. 203.[3] When used in this Act:

(1) [3a] The term "Secretary" means the Secretary of Health, Education, and Welfare; except that such term means the Secretary of the Interior with respect to problems of solid waste resulting from the extraction, proc-

---

[2] Sec. 202(b) amended by sec. 101, P.L. 91–512.
[3] Sec. 203 amended by sec. 102, P.L. 91–512.
[3a] By reason of the establishment of the U.S. Environmental Protection Agency, in December 1970, the references in the cited legislation to "The Secretary" or to "The Secretary of Health, Education and Welfare" should be *changed* to read "The Administrator" or "The Administrator, Environmental Protection Agency." Authority for this change: The President's Reorganization Plan No. 3 of 1970. There are 30 or more places in the legislation where such changes should be made, beginning with Section 203 (p. 2), entitled "Definitions."
Specific references in the legislation to the Secretary of any other Department of the Executive Branch should *not* be changed.

essing, or utilization of minerals or fossil fuels where the generation, production, or reuse of such waste is or may be controlled within the extraction, processing, or utilization facility or facilities and where such control is a feature of the technology or economy of the operation of such facility or facilities.

(2) The term "State" means a State, the District of Columbia, the Commonwealth of Puerto Rico, the Virgin Islands, Guam, and American Samoa.

(3) The term "interstate agency" means an agency of two or more municipalities in different States, or an agency established by two or more States, with authority to provide for the disposal of solid wastes and serving two or more municipalities located in different States.

(4) The term "solid waste" means garbage, refuse, and other discarded solid materials, including solid-waste materials resulting from industrial, commercial, and agricultural operations, and from community activities, but does not include solids or dissolved material in domestic sewage or other significant pollutants in water resources, such as silt, dissolved or suspended solids in industrial waste water effluents, dissolved materials in irrigation return flows or other common water pollutants.

(5) The term "solid-waste disposal" means the collection, storage, treatment, utilization, processing, or final disposal of solid waste.

(6) The term "construction," with respect to any project of construction under this act, means (A) the erection or building of new structures and acquisition of lands or interests therein, or the acquisition, replacement, expansion, remodeling, alteration, modernization, or extension of existing structures, and (B) the acquisition and installation of initial equipment of, or required in connection with, new or newly acquired structures or the expanded, remodeled, altered, modernized or extended part of existing structures (including trucks and other motor vehicles, and tractors, cranes, and other machinery) necessary for the proper utilization and operation of the facility after completion of the project; and includes preliminary planning to determine the economic and engineering feasibility and the public health and safety aspects of the project, the engineering, architectural, legal, fiscal, and economic investigations and studies, and any surveys, designs, plans, working drawings, specifications, and other action necessary for the carrying out of the project, and (C) the inspection and supervision of the process of carrying out the project to completion.

(7) the term "municipality" means a city, town, borough, county, parish, district, or other public body created by or pursuant to State law with responsibility for

the planning or administration of solid waste disposal, or an Indian tribe.

(8) The term "intermunicipal agency" means an agency established by two or more municipalities with responsibility for planning or administration of solid waste disposal.

(9) The term "recovered resources" means materials or energy recovered from solid wastes.

(10) The term "resource recovery system" means a solid waste management system which provides for collection, separation, recycling, and recovery of solid wastes, including disposal of nonrecoverable waste residues.

RESEARCH, DEMONSTRATIONS, TRAINING, AND OTHER ACTIVITIES

Research, authority of Secretary, 42 USC 3253

SEC. 204.[4] (a) The Secretary shall conduct, and encourage, cooperate with, and render financial and other assistance to appropriate public (whether Federal, State, interstate, or local) authorities, agencies, and institutions, private agencies and institutions, and individuals in the conduct of, and promote the coordination of, research, investigations, experiments, training, demonstrations, surveys, and studies relating to—

(1) any adverse health and welfare effects of the release into the environment of material present in solid waste, and methods to eliminate such effects;

(2) the operation and financing of solid waste disposal programs;

(3) the reduction of the amount of such waste and unsalvageable waste materials;

(4) the development and application of new and improved methods of collecting and disposing of solid waste and processing and recovering materials and energy from solid wastes; and

(5) the identification of solid waste components and potential materials and energy recoverable from such waste components.

(b) In carrying out the provisions of the preceding subsection, the Secretary is authorized to—

(1) collect and make available, through publications and other appropriate means, the results of, and other information pertaining to, such research and other activities, including appropriate recommendations in connection therewith;

(2) cooperate with public and private agencies, institutions, and organizations, and with any industries involved, in the preparation and the conduct of such research and other activities; and

(3) make grants-in-aid to public or private agen-

---

[4] Sec. 204 (a) amended by Sec. 103, P.L. 91–512.

85

cies and institutions and to individuals for research, training projects, surveys, and demonstrations (including construction of facilities), and provide for the conduct of research, training, surveys, and demonstrations by contract with public or private agencies and institutions and with individuals; and such contracts for research or demonstrations or both (including contracts for construction) may be made in accordance with and subject to the limitations provided with respect to research contracts of the military departments in title 10, United States Code, section 2353, except that the determination, approval, and certification required thereby shall be made by the Secretary.

70A Stat. 134

(c) Any grant, agreement, or contract made or entered into under this section shall contain provisions effective to insure that all information, uses, processes, patents and other developments resulting from any activity undertaken pursuant to such grant, agreement, or contract will be made readily available on fair and equitable terms to industries utilizing methods of solid-waste disposal and industries engaging in furnishing devices, facilities, equipment, and supplies to be used in connection with solid-waste disposal. In carrying out the provisions of this section, the Secretary and each department, agency, and officer of the Federal Government having functions or duties under this Act shall make use of and adhere to the Statement of Government Patent Policy which was promulgated by the President in his memorandum of October 10, 1963. (3 CFR, 1963 Supp., p. 238.)

SPECIAL STUDY AND DEMONSTRATION PROJECTS ON RECOVERY OF USEFUL ENERGY AND MATERIALS

Sec. 205. [5] (a) The Secretary shall carry out an investigation and study to determine—

(1) means of recovering materials and energy from solid waste, recommended uses of such materials and energy for national or international welfare, including identification of potential markets for such recovered resources, and the impact of distribution of such resources on existing markets;

(2) changes in current product characteristics and production and packaging practices which would reduce the amount of solid waste;

(3) methods of collection, separation, and containerization which will encourage efficient utilization of facilities and contribute to more effective programs of reduction, reuse, or disposal of wastes;

(4) the use of Federal procurement to develop market demand for recovered resources;

---

[5] Sec. 205 added by sec. 104(a) of P.L. 91-512.

(5) recommended incentives (including Federal grants, loans, and other assistance) and disincentives to accelerate the reclamation or recycling of materials from solid wastes, with special emphasis on motor vehicle hulks;

(6) the effect of existing public policies, including subsidies and economic incentives and disincentives, percentage depletion allowances, capital gains treatment and other tax incentives and disincentives, upon the recycling and reuse of materials, and the likely effect of the modification or elimination of such incentives and disincentives upon the reuse, recycling and conservation of such materials; and

(7) the necessity and method of imposing disposal or other charges on packaging, containers, vehicles, and other manufactured goods, which charges would reflect the cost of final disposal, the value of recoverable components of the item, and any social costs associated with nonrecycling or uncontrolled disposal of such items.

Report to President and Congress The Secretary shall from time to time, but not less frequently than annually, report the results of such investigation and study to the President and the Congress.

Demonstration projects (b) The Secretary is also authorized to carry out demonstration projects to test and demonstrate methods and techniques developed pursuant to subsection (a).

(c) Section 204 (b) and (c) shall be applicable to investigations, studies, and projects carried out under this section.

INTERSTATE AND INTERLOCAL COOPERATION

SEC. 206.[6] The Secretary shall encourage cooperative activities by the States and local governments in connection with solid-waste disposal programs; encourage where practicable, interstate, interlocal, and regional planning for, and the conduct of, interstate, interlocal, and regional solid-waste disposal programs; and encourage the enactment of improved and, so far as practicable, uniform State and local laws governing solid-waste disposal.

GRANTS FOR STATE, INTERSTATE, AND LOCAL PLANNING

SEC. 207.[7] (a) The Secretary may from time to time, upon such terms and conditions consistent with this section as he finds appropriate to carry out the purposes of this Act, make grants to State, interstate, municipal, and intermunicipal agencies, and organizations composed of

---

[6] Previous sec. 205 redesignated as sec. 206 by sec. 104(a) of P.L. 91–512.
[7] Sec. 207 added by sec. 104(b) of P.L. 91–512.

87

public officials which are eligible for assistance under section 701(g) of the Housing Act of 1954, of not to exceed 66⅔ per centum of the cost in the case of an application with respect to an area including only one municipality, and not to exceed 75 per centum of the cost in any other case, of—

82 Stat. 530
40 USC 461
Cost limitation

(1) making surveys of solid waste disposal practices and problems within the jurisdictional areas of such agencies and

(2) developing and revising solid waste disposal plans as part of regional environmental protection systems for such areas, providing for recycling or recovery of materials from wastes whenever possible and including planning for the reuse of solid waste disposal areas and studies of the effect and relationship of solid waste disposal practices on areas adjacent to waste disposal sites,

84 Stat. 1229
84 Stat. 1230

(3) developing proposals for projects to be carried out pursuant to section 208 of this Act, or

(4) planning programs for the removal and processing of abandoned motor vehicle hulks.

(b) Grants pursuant to this section may be made upon application therefor which—

(1) designates or establishes a single agency (which may be an interdepartmental agency) as the sole agency for carrying out the purposes of this section for the area involved;

(2) indicates the manner in which provision will be made to assure full consideration of all aspects of planning essential to areawide planning for proper and effective solid waste disposal consistent with the protection of the public health and welfare, including such factors as population growth, urban and metropolitan development, land use planning, water pollution control, air pollution control, and the feasibility of regional disposal and resource recovery programs;

(3) sets forth plans for expenditure of such grant, which plans provide reasonable assurance of carrying out the purposes of this section;

(4) provides for submission of such reports of the activities of the agency in carrying out the purposes of this section, in such form and containing such information, as the Secretary may from time to time find necessary for carrying out the purposes of this section and for keeping such records and affording such access thereto as he may find necessary; and

(5) provides for such fiscal-control and fund-accounting procedures as may be necessary to assure proper disbursement of and accounting for funds paid to the agency under this section.

(c) The Secretary shall make a grant under this section only if he finds that there is satisfactory assurance that the planning of solid waste disposal will be coordinated, so far as practicable, with and not duplicate other related State, interstate, regional, and local planning activities, including those financed in part with funds pursuant to section 701 of the Housing Act of 1954.

GRANTS FOR RESOURCE RECOVERY SYSTEMS AND IMPROVED SOLID WASTE DISPOSAL FACILITIES

SEC. 208.[8] (a) The Secretary is authorized to make grants pursuant to this section to any State, municipal, or interstate or intermunicipal agency for the demonstration of resource recovery systems or for the construction of new or improved solid waste disposal facilities.

(b)(1) Any grant under this section for the demonstration of a resource recovery system may be made only if it (A) is consistent with any plans which meet the requirements of section 207(b)(2) of this Act; (B) is consistent with the guidelines recommended pursuant to section 209 of this Act; (C) is designed to provide area-wide resource recovery systems consistent with the purposes of this Act, as determined by the Secretary, pursuant to regulations promulgated under subsection (d) of this section; and (D) provides an equitable system for distributing the costs associated with construction, operation, and maintenance of any resource recovery system among the users of such system.

Federal
share,
limitation

(2) The Federal share for any project to which paragraph (1) applies shall not be more than 75 percent.

(c)(1) A grant under this section for the construction of a new or improved solid waste disposal facility may be made only if—

(A) a State or interstate plan for solid waste disposal has been adopted which applies to the area involved, and the facility to be constructed (i) is consistent with such plan, (ii) is included in a comprehensive plan for the area involved which is satisfactory to the Secretary for the purposes of this Act, and (iii) is consistent with the guidelines recommended under section 209, and

(B) the project advances the state of the art by applying new and improved techniques in reducing the environmental impact of solid waste disposal, in achieving recovery of energy or resources, or in recycling useful materials.

(2) The Federal share for any project to which paragraph (1) applies shall be not more than 50 percent in

---

[8] Sec. 208 added by sec. 104(b) P.L. 91-512.

the case of a project serving an area which includes only one municipality, and not more than 75 percent in any other case.

(d) (1) The Secretary, within ninety days after the date of enactment of the Resource Recovery Act of 1970, shall promulgate regulations establishing a procedure for awarding grants under this section which— Regulations

(A) provides that projects will be carried out in communities of varying sizes, under such conditions as will assist in solving the community waste problems of urban-industrial centers, metropolitan regions, and rural areas, under representative geographic and environmental conditions; and

(B) provides deadlines for submission of, and action on, grant requests.

(2) In taking action on applications for grants under this section, consideration shall be given by the Secretary (A) to the public benefits to be derived by the construction and the propriety of Federal aid in making such grant; (B) to the extent applicable, to the economic and commercial viability of the project (including contractual arrangements with the private sector to market any resources recovered); (C) to the potential of such project for general application to community solid waste disposal problems; and (D) to the use by the applicant of comprehensive regional or metropolitan area planning.

(e) A grant under this section—

(1) may be made only in the amount of the Federal share of (A) the estimated total design and construction costs, plus (B) in the case of a grant to which subsection (b) (1) applies, the first-year operation and maintenance costs;

(2) may not be provided for land acquisition or (except as otherwise provided in paragraph (1) (B)) for operating or maintenance costs;

(3) may not be made until the applicant has made provision satisfactory to the Secretary for proper and efficient operation and maintenance of the project (subject to paragraph (1) (B)); and

(4) may be made subject to such conditions and requirements, in addition to those provided in this section, as the Secretary may require to properly carry out his functions pursuant to this Act.

For purposes of paragraph (1), the non-Federal share may be in any form, including, but not limited to, lands or interests therein needed for the project or personal property or services, the value of which shall be determined by the Secretary.

(f) (1) Not more than 15 percent of the total of funds authorized to be appropriated under section 216(a) (3) for any fiscal year to carry out this section shall be granted under this section for projects in any one State. Limitation

(2) The Secretary shall prescribe by regulation the manner in which this subsection shall apply to a grant under this section for a project in an area which includes all or part of more than one State.

RECOMMENDED GUIDELINES

Sec. 209.[9] (a) The Secretary shall, in cooperation with appropriate State, Federal, interstate, regional, and local agencies, allowing for public comment by other interested parties, as soon as practicable after the enactment of the Resource Recovery Act of 1970, recommend to appropriate agencies and publish in the Federal Register guidelines for solid waste recovery, collection, separation, and disposal systems (including systems for private use), which shall be consistent with public health and welfare, and air and water quality standards and adaptable to appropriate land-use plans. Such guidelines shall apply to such systems whether on land or water and shall be revised from time to time.

(b) (1) The Secretary shall, as soon as practicable, recommend model codes, ordinances, and statutes which are designed to implement this section and the purposes of this Act.

(2) The Secretary shall issue to appropriate Federal, interstate, regional, and local agencies information on technically feasible solid waste collection, separation, disposal, recycling, and recovery methods, including data on the cost of construction, operation, and maintenance of such methods.

GRANTS OR CONTRACTS FOR TRAINING PROJECTS

Sec. 210.[10] (a) The Secretary is authorized to make grants to, and contracts with, any eligible organization. For purposes of this section the term "eligible organization" means a State or interstate agency, a municipality, educational institution, and any other organization which is capable of effectively carrying out a project which may be funded by grant under subsection (b) of this section.

(b) (1) Subject to the provisions of paragraph (2), grants or contracts may be made to pay all or a part of the costs, as may be determined by the Secretary, of any project operated or to be operated by an eligible organization, which is designed—

(A) to develop, expand, or carry out a program (which may combine training, education, and employment) for training persons for occupations involving the management, supervision, design, op-

---

[9] Sec. 209 added by sec. 104 (b) P.L. 91–512.
[10] Sec. 210 added by sec. 104 (b) P.L. 91–512.

eration, or maintenance of solid waste disposal and resources recovery equipment and facilities; or

(B) to train instructors and supervisory personnel to train or supervise persons in occupations involving the design, operation, and maintenance of solid waste disposal and resource recovery equipment and facilities.

(2) A grant or contract authorized by paragraph (1) of this subsection may be made only upon application to the Secretary at such time or times and containing such information as he may prescribe, except that no such application shall be approved unless it provides for the same procedures and reports (and access to such reports and to other records) as is required by section 207(b) (4) and (5) with respect to applications made under such section.

(c) The Secretary shall make a complete investigation and study to determine— Study

(1) the need for additional trained State and local personnel to carry out plans assisted under this Act and other solid waste and resource recovery programs;

(2) means of using existing training programs to train such personnel; and

(3) the extent and nature of obstacles to employment and occupational advancement in the solid waste disposal and resource recovery field which may limit either available manpower or the advancement of personnel in such field.

He shall report the results of such investigation and study, including his recommendations to the President and the Congress not later than one year after enactment of this Act. Report to President and Congress

APPLICABILITY OF SOLID WASTE DISPOSAL GUIDELINES TO EXECUTIVE AGENCIES

SEC. 211.[11] (a) (1) If—

(A) an Executive agency (as defined in section 105 of title 5, United States Code) has jurisdiction over any real property or facility the operation or administration of which involves such agency in solid waste disposal activities, or

(B) such an agency enters into a contract with any person for the operation by such person of any Federal property or facility, and the performance of such contract involves such person in solid waste disposal activities,

then such agency shall insure compliance with the guidelines recommended under section 209 and the purposes Compliance.

---

[11] Sec. 211 added by sec. 104(b) P.L. 91–512.

of this Act in the operation or administration of such property or facility, or the performance of such contract, as the case may be.

(2) Each Executive agency which conducts any activity—

(A) which generates solid waste, and

(B) which, if conducted by a person other than such agency, would require a permit or license from such agency in order to dispose of such solid waste,

shall insure compliance with such guidelines and the purposes of this Act in conducting such activity.

(3) Each Executive agency which permits the use of Federal property for purposes of disposal of solid waste shall insure compliance with such guidelines and the purposes of this Act in the disposal of such waste.

(4) The President shall prescribe regulations to carry out this subsection.

Presidential regulations.

(b) Each Executive agency which issues any license or permit for disposal of solid waste shall, prior to the issuance of such license or permit, consult with the Secretary to insure compliance with guidelines recommended under section 209 and the purposes of this Act.

NATIONAL DISPOSAL SITES STUDY

Report to Congress.

SEC. 212.[12] The Secretary shall submit to the Congress no later than two years after the date of enactment of the Resource Recovery Act of 1970, a comprehensive report and plan for the creation of a system of national disposal sites for the storage and disposal of hazardous wastes, including radioactive, toxic chemical, biological, and other wastes which may endanger public health or welfare. Such report shall include: (1) a list of materials which should be subject to disposal in any such site; (2) current methods of disposal of such materials; (3) recommended methods of reduction, neutralization, recovery, or disposal of such materials; (4) an inventory of possible sites including existing land or water disposal sites operated or licensed by Federal agencies; (5) an estimate of the cost of developing and maintaining sites including consideration of means for distributing the short- and long-term costs of operating such sites among the users thereof; and (6) such other information as may be appropriate.

LABOR STANDARDS

40 U.S.C. 461

SEC. 213.[13] No grant for a project of construction under this Act shall be made unless the Secretary finds that the application contains or is supported by reason-

[12] Sec. 212 added by sec. 104(b) of P.L. 91–512.
[13] Former secs. 207 through 210 redesignated as secs. 213 through 216 by sec. 104(b) of P.L. 91–512.

able assurance that all laborers and mechanics employed by contractors or subcontractors on projects of the type covered by the Davis-Bacon Act, as amended (40 U.S.C. 276a—276a–5), will be paid wages at rates not less than those prevailing on similar work in the locality as determined by the Secretary of Labor in accordance with that Act; and the Secretary of Labor shall have with respect to the labor standards specified in this section the authority and functions set forth in Reorganization Plan Numbered 14 of 1950 (15 F.R. 3176; 5 U.S.C. 133z–15) and section 2 of the Act of June 13, 1934, as amended (40 U.S.C. 276c). 49 Stat. 1101; 78 Stat. 238 63 Stat. 108 64 Stat. 1267

OTHER AUTHORITY NOT AFFECTED

SEC. 214. This Act shall not be construed as superseding or limiting the authorities and responsibilities, under any other provisions of law, of the Secretary of Health, Education, and Welfare, the Secretary of the Interior, or any other Federal officer, department, or agency.

GENERAL PROVISIONS

SEC. 215.[14] (a) Payments of grants under this Act may be made (after necessary adjustment on account of previously made underpayments or overpayments) in advance or by way of reimbursement, and in such installments and on such conditions as the Secretary may determine.

(b) No grant may be made under this Act to any private profitmaking organization. Grants, prohibition.

SEC. 216.[15] (a) (1) There are authorized to be appropriated to the Secretary of Health, Education, and Welfare for carrying out the provisions of this Act (including, but not limited to, section 208), not to exceed $41,500,000 for the fiscal year ending June 30, 1971. Appropriation.

(2) There are authorized to be appropriated to the Administrator of the Environmental Protection Agency to carry out the provisions of this Act, other than section 208, not to exceed $72,000,000 for the fiscal year ending June 30, 1972, and not to exceed $76,000,000 for the fiscal year ending June 30, 1973, and not to exceed $76,000,000 for the fiscal year ending June 30, 1974.[16]

(3) There are authorized to be appropriated to the Administrator of the Environmental Protection Agency to carry out section 208 of this Act not to exceed $80,000,000 for the fiscal year ending June 30, 1972, and not to exceed $140,000,000 for the fiscal year ending June 30, 1973, and not to exceed $140,000,000 for the fiscal year ending June 30, 1974.[16]

[14] Sec. 215 as redesignated by sec. 104(b) of P.L. 91–512 further amended by sec. 104(c) of that Act.
[15] Sec. 216 as redesignated by sec. 104(b) of P.L. 91–512 further amended by sec. 105 of that Act.
[16] P.L. 93-14 extended authorization of funding to June 30, 1974.

(b) There are authorized to be appropriated to the Secretary of the Interior to carry out this Act not to exceed $8,750,000 for the fiscal year ending June 30, 1971, not to exceed $20,000,000 for the fiscal year ending June 30, 1972, not to exceed $22,500,000 for the fiscal year ending June 30, 1973, and not to exceed $22,500,000 for the fiscal year ending June 30, 1974.[16] Prior to expending any funds authorized to be appropriated by this subsection, the Secretary of the Interior shall consult with the Secretary of Health, Education, and Welfare to assure that the expenditure of such funds will be consistent with the purposes of this Act.

Program Evaluation. (c) Such portion as the Secretary may determine, but not more than 1 per centum, of any appropriation for grants, contracts, or other payments under any provision of this Act for any fiscal year beginning after June 30, 1970, shall be available for evaluation (directly, or by grants or contracts) of any program authorized by this Act.

Funds. availability. (d) Sums appropriated under this section shall remain available until expended.

Summary

The Solid Waste Disposal Act was enacted to promote the demonstration, construction, and application of solid waste management and recovery systems which preserve and enhance the quality of air, water, and land resources. In addition, the Act also provides technical and financial assistance to States and local governments and interstate agencies in the planning and development of resource recovery and solid waste disposal programs. The Act has provisions designed to promote a national research and development program for improved management techniques, more effective organizational arrangements, and new and improved methods of collection, separation, recovery, and recycling of solid wastes, and the environmentally safe disposal of nonrecoverable residues. Provisions in the Act provide for the promulgation of guidelines for solid waste collection, transport, separation, recovery, and disposal systems, as well as training grants in occupations involving the design, operation, and maintenance of solid waste disposal systems.

Review Questions

1. What are two specific purposes of the Solid Waste Disposal Act?

2. Identify one economic factor that aided the passage of the Solid Waste Disposal Act?

3. Identify one political factor that aided the passage of the Solid Waste Disposal Act?

4. Identify one social factor that aided the passage of the Solid Waste Disposal Act?

5. What effect has the Solid Waste Disposal Act and the Resource Recovery Act of 1970 had on the recycling and recovery of valuable waste materials with regard to new products?

6. What are the strengths of the Solid Waste Disposal Act with regard to pollution control problems?

7. What are the weaknesses of the Solid Waste Disposal Act with regard to current pollution control problems?

8. What are the strengths of the Solid Waste Disposal Act with regard to current energy conservation measures?

9. What are the weaknesses of the Solid Waste Disposal Act with regard to current energy conservation measures?

noise

OBJECTIVES:

The student should be able to identify the specific purposes of the Noise Control Act of 1972.

The student should be able to describe those economic, political, and social factors leading to the passage of the Noise Control Act of 1972.

The student should be able to describe the effect the Noise Control Act of 1972 has had on various types of industries.

The student should be able to identify the strengths and weaknesses of the Noise Control Act of 1972 as they relate to current pollution control problems.

INTRODUCTION:

In the United States, we are beginning to realize that man should not tolerate indefinitely the increasing noise that presently characterizes the modern, industrialized nation. Mechanically-generated noise—from the jet plane, the power mower, the diesel truck, the motorcycle, the jackhammer—is a profound annoyance to most people. It has increased dramatically in volume in the last 30 years and continues to rise in urban areas at a rate estimated at one decibel per annum.

It has been clearly demonstrated that workers in certain occupations suffer noise-induced hearing loss. The effects of community noise on hearing are not yet known. However some 20 percent of the United States population, in addition to those exposed to excessive occupational noise, suffer measurable hearing impairment by their fifties, whereas people in non-industrial societies

TOWARD A NEW ENVIRONMENTAL ETHIC, U.S. Government Printing Office: 1971-O — 443-062, p. 24.

experience no such loss.

Hearing loss is not the only potential health problem associated with noise. Evidence is growing that intense noise may affect other psychologic and physiologic functions of man.

We have tended in the past to accept noise as a phenomenon essentially beyond control. As a result, we have failed to take full advantage of the many noise suppression techniques that are available.

The technology to curb noise from construction equipment, railroad equipment, cars, trucks, and buses exists today. Much can be done to reduce noise associated with aircraft. Auto tires can be made with non-squeal treads. Silence can be designed into machinery for the home, office and factory at reasonable cost. And there is no mystery about constructing sound-proof buildings of all kinds.

In accordance with the Noise Abatement and Control Act of 1970:

• EPA has set up an office of Noise Abatement and Control to evaluate health hazards to the extent possible, summarize the state-of-the-art in noise suppression technology, and recommend a program of counter-measures to Congress not later than December 31, 1971.

• Public hearings have been held in various parts of the country to determine the extent of the problem and identify required control measures.

Public Law 92-574
92nd Congress, H. R. 11021
October 27, 1972

An Act

To control the emission of noise detrimental to the human environment, and for other purposes.

Be it enacted by the Senate and House of Representatives of the United States of America in Congress assembled,

Noise Control Act of 1972.

SHORT TITLE

SECTION 1. This Act may be cited as the "Noise Control Act of 1972".

FINDINGS AND POLICY

SEC. 2. (a) The Congress finds—

(1) that inadequately controlled noise presents a growing danger to the health and welfare of the Nation's population, particularly in urban areas;

(2) that the major sources of noise include transportation vehicles and equipment, machinery, appliances, and other products in commerce; and

(3) that, while primary responsibility for control of noise rests with State and local governments, Federal action is essential to deal with major noise sources in commerce control of which require national uniformity of treatment.

(b) The Congress declares that it is the policy of the United States to promote an environment for all Americans free from noise that jeopardizes their health or welfare. To that end, it is the purpose of this Act to establish a means for effective coordination of Federal research and activities in noise control, to authorize the establishment of Federal noise emission standards for products distributed in commerce, and to provide information to the public respecting the noise emission and noise reduction characteristics of such products.

DEFINITIONS

SEC. 3. For purposes of this Act:

(1) The term "Administrator" means the Administrator of the Environmental Protection Agency.

(2) The term "person" means an individual, corporation, partnership, or association, and (except as provided in sections 11(e) and 12(a)) includes any officer, employee, department, agency, or instrumentality of the United States, a State, or any political subdivision of a State.

(3) The term "product" means any manufactured article or goods or component thereof; except that such term does not include—

(A) any aircraft, aircraft engine, propeller, or appliance, as such terms are defined in section 101 of the Federal Aviation Act of 1958; or

72 Stat. 737.
49 USC 1301.

(B)(i) any military weapons or equipment which are designed for combat use; (ii) any rockets or equipment which are designed for research, experimental, or developmental work to be performed by the National Aeronautics and Space Administration; or (iii) to the extent provided by regulations of the Administrator, any other machinery or equipment designed for use in experimental work done by or for the Federal Government.

(4) The term "ultimate purchaser" means the first person who in good faith purchases a product for purposes other than resale.

NOISE CONTROL ACT, October 27, 1972, Public Law 92-574, 86 STAT. 1234-1250.

99

(5) The term "new product" means (A) a product the equitable or legal title of which has never been transferred to an ultimate purchaser, or (B) a product which is imported or offered for importation into the United States and which is manufactured after the effective date of a regulation under section 6 or section 8 which would have been applicable to such product had it been manufactured in the United States.

(6) The term "manufacturer" means any person engaged in the manufacturing or assembling of new products, or the importing of new products for resale, or who acts for, and is controlled by, any such person in connection with the distribution of such products.

(7) The term "commerce" means trade, traffic, commerce, or transportation—

(A) between a place in a State and any place outside thereof, or

(B) which affects trade, traffic, commerce, or transportation described in subparagraph (A).

(8) The term "distribute in commerce" means sell in, offer for sale in, or introduce or deliver for introduction into, commerce.

(9) The term "State" includes the District of Columbia, the Commonwealth of Puerto Rico, the Virgin Islands, American Samoa, Guam, and the Trust Territory of the Pacific Islands.

80 Stat. 379.

(10) The term "Federal agency" means an executive agency (as defined in section 105 of title 5, United States Code) and includes the United States Postal Service.

(11) The term "environmental noise" means the intensity, duration, and the character of sounds from all sources.

FEDERAL PROGRAMS

Sec. 4. (a) The Congress authorizes and directs that Federal agencies shall, to the fullest extent consistent with their authority under Federal laws administered by them, carry out the programs within their control in such a manner as to further the policy declared in section 2(b).

(b) Each department, agency, or instrumentality of the executive, legislative, and judicial branches of the Federal Government—

(1) having jurisdiction over any property or facility, or

(2) engaged in any activity resulting, or which may result, in the emission of noise,

shall comply with Federal, State, interstate, and local requirements respecting control and abatement of environmental noise to the same

Compliance exemption, Presidential authority.

extent that any person is subject to such requirements. The President may exempt any single activity or facility, including noise emission sources or classes thereof, of any department, agency, or instrumentality in the executive branch from compliance with any such requirement if he determines it to be in the paramount interest of the United States to do so; except that no exemption, other than for those products referred to in section 3(3)(B) of this Act, may be granted from the requirements of sections 6, 17, and 18 of this Act. No such exemption shall be granted due to lack of appropriation unless the President shall have specifically requested such appropriation as a part of the budgetary process and the Congress shall have failed to make available such requested appropriation. Any exemption shall be for a period not in excess of one year, but additional exemptions may be granted for periods of not to exceed one year upon the

Report to Congress.

President's making a new determination. The President shall report each January to the Congress all exemptions from the requirements of this section granted during the preceding calendar year, together with his reason for granting such exemption.

(c)(1) The Administrator shall coordinate the programs of all Federal agencies relating to noise research and noise control. Each Federal agency shall, upon request, furnish to the Administrator such information as he may reasonably require to determine the nature, scope, and results of the noise-research and noise-control programs of the agency.

(2) Each Federal agency shall consult with the Administrator in prescribing standards or regulations respecting noise. If at any time the Administrator has reason to believe that a standard or regulation, or any proposed standard or regulation, of any Federal agency respecting noise does not protect the public health and welfare to the extent he believes to be required and feasible, he may request such agency to review and report to him on the advisability of revising such standard or regulation to provide such protection. Any such request may be published in the Federal Register and shall be accompanied by a detailed statement of the information on which it is based. Such agency shall complete the requested review and report to the Administrator within such time as the Administrator specifies in the request, but such time specified may not be less than ninety days from the date the request was made. The report shall be published in the Federal Register and shall be accompanied by a detailed statement of the findings and conclusions of the agency respecting the revision of its standard or regulation. With respect to the Federal Aviation Administration, section 611 of the Federal Aviation Act of 1958 (as amended by section 7 of this Act) shall apply in lieu of this paragraph.

(3) On the basis of regular consultation with appropriate Federal agencies, the Administrator shall compile and publish, from time to time, a report on the status and progress of Federal activities relating to noise research and noise control. This report shall describe the noise-control programs of each Federal agency and assess the contributions of those programs to the Federal Government's overall efforts to control noise.

IDENTIFICATION OF MAJOR NOISE SOURCES ; NOISE CRITERIA AND CONTROL
TECHNOLOGY

SEC. 5. (a)(1) The Administrator shall, after consultation with appropriate Federal agencies and within nine months of the date of the enactment of this Act, develop and publish criteria with respect to noise. Such criteria shall reflect the scientific knowledge most useful in indicating the kind and extent of all identifiable effects on the public health or welfare which may be expected from differing quantities and qualities of noise.

(2) The Administrator shall, after consultation with appropriate Federal agencies and within twelve months of the date of the enactment of this Act, publish information on the levels of environmental noise the attainment and maintenance of which in defined areas under various conditions are requisite to protect the public health and welfare with an adequate margin of safety.

(b) The Administrator shall, after consultation with appropriate Federal agencies, compile and publish a report or series of reports (1) identifying products (or classes of products) which in his judgment are major sources of noise, and (2) giving information on techniques for control of noise from such products, including available data on the technology, costs, and alternative methods of noise control. The first such report shall be published not later than eighteen months after the date of enactment of this Act.

(c) The Administrator shall 'from time to time review and, as appropriate, revise or supplement any criteria or reports published under this section.

(d) Any report (or revision thereof) under subsection (b) (1) identifying major noise sources shall be published in the Federal Register. The publication or revision under this section of any criteria or information on control techniques shall be announced in the Federal Register, and copies shall be made available to the general public.

NOISE EMISSION STANDARDS FOR PRODUCTS DISTRIBUTED IN COMMERCE

SEC. 6. (a) (1) The Administrator shall publish proposed regulations, meeting the requirements of subsection (c), for each product—
    (A) which is identified (or is part of a class identified) in any report published under section 5(b) (1) as a major source of noise,
    (B) for which, in his judgment, noise emission standards are feasible, and
    (C) which falls in one of the following categories:
        (i) Construction equipment.
        (ii) Transportation equipment (including recreational vehicles and related equipment).
        (iii) Any motor or engine (including any equipment of which an engine or motor is an integral part).
        (iv) Electrical or electronic equipment.

(2) (A) Initial proposed regulations under paragraph (1) shall be published not later than eighteen months after the date of enactment of this Act, and shall apply to any product described in paragraph (1) which is identified (or is a part of a class identified) as a major source of noise in any report published under section 5(b) (1) on or before the date of publication of such initial proposed regulations.

(B) In the case of any product described in paragraph (1) which is identified (or is part of a class identified) as a major source of noise in a report published under section 5(b) (1) after publication of the initial proposed regulations under subparagraph (A) of this paragraph, regulations under paragraph (1) for such product shall be proposed and published by the Administrator not later than eighteen months after such report is published.

(3) After proposed regulations respecting a product have been published under paragraph (2), the Administrator shall, unless in his judgment noise emission standards are not feasible for such product, prescribe regulations, meeting the requirements of subsection (c), for such product—
    (A) not earlier than six months after publication of such proposed regulations, and
    (B) not later than—
        (i) twenty-four months after the date of enactment of this Act, in the case of a product subject to proposed regulations published under paragraph (2) (A), or
        (ii) in the case of any other product, twenty-four months after the publication of the report under section 5(b) (1) identifying it (or a class of products of which it is a part) as a major source of noise.

(b) The Administrator may publish proposed regulations, meeting the requirements of subsection (c), for any product for which he is not required by subsection (a) to prescribe regulations but for which, in his judgment, noise emission standards are feasible and are requisite to protect the public health and welfare. Not earlier than six months after the date of publication of such proposed regulations respecting such product, he may prescribe regulations, meeting the requirements of subsection (c), for such product.

(c)(1) Any regulation prescribed under subsection (a) or (b) of this section (and any revision thereof) respecting a product shall include a noise emission standard which shall set limits on noise emissions from such product and shall be a standard which in the Administrator's judgment, based on criteria published under section 5, is requisite to protect the public health and welfare, taking into account the magnitude and conditions of use of such product (alone or in combination with other noise sources), the degree of noise reduction achievable through the application of the best available technology, and the cost of compliance. In establishing such a standard for any product, the Administrator shall give appropriate consideration to standards under other laws designed to safeguard the health and welfare of persons, including any standards under the National Traffic and Motor Vehicle Safety Act of 1966, the Clean Air Act, and the Federal Water Pollution Control Act. Any such noise emission standards shall be a performance standard. In addition, any regulation under subsection (a) or (b) (and any revision thereof) may contain testing procedures necessary to assure compliance with the emission standard in such regulation, and may contain provisions respecting instructions of the manufacturer for the maintenance, use, or repair of the product.

80 Stat. 718.
15 USC 1381 note.
81 Stat. 485.
42 USC 1857 note.
Ante, p. 816.

(2) After publication of any proposed regulations under this section, the Administrator shall allow interested persons an opportunity to participate in rulemaking in accordance with the first sentence of section 553(c) of title 5, United States Code.

80 Stat. 383.

(3) The Administrator may revise any regulation prescribed by him under this section by (A) publication of proposed revised regulations, and (B) the promulgation, not earlier than six months after the date of such publication, of regulations making the revision; except that a revision which makes only technical or clerical corrections in a regulation under this section may be promulgated earlier than six months after such date if the Administrator finds that such earlier promulgation is in the public interest.

(d)(1) On and after the effective date of any regulation prescribed under subsection (a) or (b) of this section, the manufacturer of each new product to which such regulation applies shall warrant to the ultimate purchaser and each subsequent purchaser that such product is designed, built, and equipped so as to conform at the time of sale with such regulation.

(2) Any cost obligation of any dealer incurred as a result of any requirement imposed by paragraph (1) of this subsection shall be borne by the manufacturer. The transfer of any such cost obligation from a manufacturer to any dealer through franchise or other agreement is prohibited.

Cost obligations, transfer prohibition.

(3) If a manufacturer includes in any advertisement a statement respecting the cost or value of noise emission control devices or systems, such manufacturer shall set forth in such statement the cost or value attributed to such devices or systems by the Secretary of Labor (through the Bureau of Labor Statistics). The Secretary of Labor, and his representatives, shall have the same access for this purpose to the books, documents, papers, and records of a manufacturer as the Comptroller General has to those of a recipient of assistance for purposes of section 311 of the Clean Air Act.

81 Stat. 505;
84 Stat. 1705.
42 USC 1857j.
Prohibitions.

(e)(1) No State or political subdivision thereof may adopt or enforce—

(A) with respect to any new product for which a regulation has been prescribed by the Administrator under this section, any law or regulation which sets a limit on noise emissions from such new product and which is not identical to such regulation of the Administrator; or

(B) with respect to any component incorporated into such new product by the manufacturer of such product, any law or regulation setting a limit on noise emissions from such component when so incorporated.

(2) Subject to sections 17 and 18, nothing in this section precludes or denies the right of any State or political subdivision thereof to establish and enforce controls on environmental noise (or one or more sources thereof) through the licensing, regulation, or restriction of the use, operation, or movement of any product or combination of products.

AIRCRAFT NOISE STANDARDS

SEC. 7. (a) The Administrator, after consultation with appropriate Federal, State, and local agencies and interested persons, shall conduct a study of the (1) adequacy of Federal Aviation Administration flight and operational noise controls; (2) adequacy of noise emission standards on new and existing aircraft, together with recommendations on the retrofitting and phaseout of existing aircraft; (3) implications of identifying and achieving levels of cumulative noise exposure around airports; and (4) additional measures available to airport operators and local governments to control aircraft noise. He shall report on such study to the Committee on Interstate and Foreign Commerce of the House of Representatives and the Committees on Commerce and Public Works of the Senate within nine months after the date of the enactment of this Act.

82 Stat. 395. (b) Section 611 of the Federal Aviation Act of 1958 (49 U.S.C. 1431) is amended to read as follows:

"CONTROL AND ABATEMENT OF AIRCRAFT NOISE AND SONIC BOOM

"SEC. 611. (a) For purposes of this section:

"FAA." "(1) The term 'FAA' means Administrator of the Federal Aviation Administration.

"EPA." "(2) The term 'EPA' means the Administrator of the Environmental Protection Agency.

Standards and regulations. "(b)(1) In order to afford present and future relief and protection to the public health and welfare from aircraft noise and sonic boom, the FAA, after consultation with the Secretary of Transportation and with EPA, shall prescribe and amend standards for the measurement of aircraft noise and sonic boom and shall prescribe and amend such regulations as the FAA may find necessary to provide for the control and abatement of aircraft noise and sonic boom, including the application of such standards and regulations in the issuance, amendment, modification, suspension, or revocation of any certificate authorized by this title. No exemption with respect to any standard or regulation under this section may be granted under any provision of this Act unless the FAA shall have consulted with EPA before such exemption is granted, except that if the FAA determines that safety in air commerce or air transportation requires that such an exemption be granted before EPA can be consulted, the FAA shall consult with EPA as soon as practicable after the exemption is granted.

72 Stat. 776.
49 USC. 1423. "(2) The FAA shall not issue an original type certificate under section 603(a) of this Act for any aircraft for which substantial noise abatement can be achieved by prescribing standards and regulations in accordance with this section, unless he shall have prescribed standards and regulations in accordance with this section which apply to such aircraft and which protect the public from aircraft noise and sonic boom, consistent with the considerations listed in subsection (d).

"(c)(1) Not earlier than the date of submission of the report required by section 7(a) of the Noise Control Act of 1972, EPA shall submit to the FAA proposed regulations to provide such control and abatement of aircraft noise and sonic boom (including control and abatement through the exercise of any of the FAA's regulatory authority over air commerce or transportation or over aircraft or airport operations) as EPA determines is necessary to protect the public health and welfare. The FAA shall consider such proposed regulations submitted by EPA under this paragraph and shall, within thirty days of the date of its submission to the FAA, publish the proposed regulations in a notice of proposed rulemaking. Within sixty days after such publication, the FAA shall commence a hearing at which interested persons shall be afforded an opportunity for oral (as well as written) presentations of data, views, and arguments. Within a reasonable time after the conclusion of such hearing and after consultation with EPA, the FAA shall—

Proposed regulations, submittal to FAA.

Publication.

Hearing.

"(A) in accordance with subsection (b), prescribe regulations (i) substantially as they were submitted by EPA, or (ii) which are a modification of the proposed regulations submitted by EPA, or

"(B) publish in the Federal Register a notice that it is not prescribing any regulation in response to EPA's submission of proposed regulations, together with a detailed explanation providing reasons for the decision not to prescribe such regulations.

Publication in Federal Register.

"(2) If EPA has reason to believe that the FAA's action with respect to a regulation proposed by EPA under paragraph (1)(A)(ii) or (1)(B) of this subsection does not protect the public health and welfare from aircraft noise or sonic boom, consistent with the considerations listed in subsection (d) of this section, EPA shall consult with the FAA and may request the FAA to review, and report to EPA on, the advisability of prescribing the regulation originally proposed by EPA. Any such request shall be published in the Federal Register and shall include a detailed statement of the information on which it is based. The FAA shall complete the review requested and shall report to EPA within such time as EPA specifies in the request, but such time specified may not be less than ninety days from the date the request was made. The FAA's report shall be accompanied by a detailed statement of the FAA's findings and the reasons for the FAA's conclusions; shall identify any statement filed pursuant to section 102(2)(C) of the National Environmental Policy Act of 1969 with respect to such action of the FAA under paragraph (1) of this subsection; and shall specify whether (and where) such statements are available for public inspection. The FAA's report shall be published in the Federal Register, except in a case in which EPA's request proposed specific action to be taken by the FAA, and the FAA's report indicates such action will be taken.

Report request, publication in Federal Register.

83 Stat. 853. 42 USC 4332.

Report, publication in Federal Register.

"(3) If, in the case of a matter described in paragraph (2) of this subsection with respect to which no statement is required to be filed under such section 102(2)(C), the report of the FAA indicates that the proposed regulation originally submitted by EPA should not be made, then EPA may request the FAA to file a supplemental report, which shall be published in the Federal Register within such a period as EPA may specify (but such time specified shall not be less than ninety days from the date the request was made), and which shall contain a comparison of (A) the environmental effects (including those which cannot be avoided) of the action actually taken by the FAA in response to EPA's proposed regulations, and (B) EPA's proposed regulations.

Supplemental report, publication in Federal Register.

"(d) In prescribing and amending standards and regulations under this section, the FAA shall—

"(1) consider relevant available data relating to aircraft noise and sonic boom, including the results of research, development, testing, and evaluation activities conducted pursuant to this Act and the Department of Transportation Act;

80 Stat. 931.
49 USC 1651
note.

"(2) consult with such Federal, State, and interstate agencies as he deems appropriate;

"(3) consider whether any proposed standard or regulation is consistent with the highest degree of safety in air commerce or air transportation in the public interest;

"(4) consider whether any proposed standard or regulation is economically reasonable, technologically practicable, and appropriate for the particular type of aircraft, aircraft engine, appliance, or certificate to which it will apply; and

"(5) consider the extent to which such standard or regulation will contribute to carrying out the purposes of this section.

Notice and
appeal.

72 Stat. 779;
85 Stat. 481.
49 USC 1429.

"(e) In any action to amend, modify, suspend, or revoke a certificate in which violation of aircraft noise or sonic boom standards or regulations is at issue, the certificate holder shall have the same notice and appeal rights as are contained in section 609, and in any appeal to the National Transportation Safety Board, the Board may amend, modify, or reverse the order of the FAA if it finds that control or abatement of aircraft noise or sonic boom and the public health and welfare do not require the affirmation of such order, or that such order is not consistent with safety in air commerce or air transportation."

(c) All—

(1) standards, rules, and regulations prescribed under section 611 of the Federal Aviation Act of 1958, and

(2) exemptions, granted under any provision of the Federal Aviation Act of 1958, with respect to such standards, rules, and regulations,

72 Stat. 731.
49 USC 1301
note.

which are in effect on the date of the enactment of this Act, shall continue in effect according to their terms until modified, terminated, superseded, set aside, or repealed by the Administrator of the Federal Aviation Administration in the exercise of any authority vested in him, by a court of competent jurisdiction, or by operation of law.

LABELING

Regulations.

SEC. 8. (a) The Administrator shall by regulation designate any product (or class thereof)—

(1) which emits noise capable of adversely affecting the public health or welfare; or

(2) which is sold wholly or in part on the basis of its effectiveness in reducing noise.

(b) For each product (or class thereof) designated under subsection (a) the Administrator shall by regulation require that notice be given to the prospective user of the level of the noise the product emits, or of its effectiveness in reducing noise, as the case may be. Such regulations shall specify (1) whether such notice shall be affixed to the product or to the outside of its container, or to both, at the time of its sale to the ultimate purchaser or whether such notice shall be given to the prospective user in some other manner, (2) the form of the notice, and (3) the methods and units of measurement to be used. Sections 6(c)(2) shall apply to the prescribing of any regulation under this section.

(c) This section does not prevent any State or political subdivision thereof from regulating product labeling or information respecting products in any way not in conflict with regulations prescribed by the Administrator under this section.

106

SEC. 9. The Secretary of the Treasury shall, in consultation with the Administrator, issue regulations to carry out the provisions of this Act with respect to new products imported or offered for importation. *Regulations.*

PROHIBITED ACTS

SEC. 10. (a) Except as otherwise provided in subsection (b), the following acts or the causing thereof are prohibited:

(1) In the case of a manufacturer, to distribute in commerce any new product manufactured after the effective date of a regulation prescribed under section 6 which is applicable to such product, except in conformity with such regulation.

(2) (A) The removal or rendering inoperative by any person, other than for purposes of maintenance, repair, or replacement, of any device or element of design incorporated into any product in compliance with regulations under section 6, prior to its sale or delivery to the ultimate purchaser or while it is in use, or (B) the use of a product after such device or element of design has been removed or rendered inoperative by any person.

(3) In the case of a manufacturer, to distribute in commerce any new product manufactured after the effective date of a regulation prescribed under section 8(b) (requiring information respecting noise) which is applicable to such product, except in conformity with such regulation.

(4) The removal by any person of any notice affixed to a product or container pursuant to regulations prescribed under section 8(b), prior to sale of the product to the ultimate purchaser.

(5) The importation into the United States by any person of any new product in violation of a regulation prescribed under section 9 which is applicable to such product.

(6) The failure or refusal by any person to comply with any requirement of section 11(d) or 13(a) or regulations prescribed under section 13(a), 17, or 18.

(b) (1) For the purpose of research, investigations, studies, demonstrations, or training, or for reasons of national security, the Administrator may exempt for a specified period of time any product, or class thereof, from paragraphs (1), (2), (3), and (5) of subsection (a), upon such terms and conditions as he may find necessary to protect the public health or welfare. *Exemptions.*

(2) Paragraphs (1), (2), (3), and (4) of subsection (a) shall not apply with respect to any product which is manufactured solely for use outside any State and which (and the container of which) is labeled or otherwise marked to show that it is manufactured solely for use outside any State; except that such paragraphs shall apply to such product if it is in fact distributed in commerce for use in any State.

ENFORCEMENT

SEC. 11. (a) Any person who willfully or knowingly violates paragraph (1), (3), (5), or (6) of subsection (a) of section 10 of this Act shall be punished by a fine of not more than $25,000 per day of violation, or by imprisonment for not more than one year, or by both. If the conviction is for a violation committed after a first conviction of such person under this subsection, punishment shall be by a fine of not more than $50,000 per day of violation, or by imprisonment for not more than two years, or by both. *Penalty.*

(b) For the purpose of this section, each day of violation of any paragraph of section 10(a) shall constitute a separate violation of that section.

Jurisdiction.
(c) The district courts of the United States shall have jurisdiction of actions brought by and in the name of the United States to restrain any violations of section 10(a) of this Act.

(d)(1) Whenever any person is in violation of section 10(a) of this Act, the Administrator may issue an order specifying such relief as he determines is necessary to protect the public health and welfare.

(2) Any order under this subsection shall be issued only after notice and opportunity for a hearing in accordance with section 554 of title 5 of the United States Code.

80 Stat. 384.
"Person."
(e) The term "person," as used in this section, does not include a department, agency, or instrumentality of the United States.

CITIZEN SUITS

SEC. 12. (a) Except as provided in subsection (b), any person (other than the United States) may commence a civil action on his own behalf—

(1) against any person (including (A) the United States, and (B) any other governmental instrumentality or agency to the

USC prec.
title 1.
extent permitted by the eleventh amendment to the Constitution) who is alleged to be in violation of any noise control requirement (as defined in subsection (e)), or

(2) against—

(A) the Administrator of the Environmental Protection Agency where there is alleged a failure of such Administrator to perform any act or duty under this Act which is not discretionary with such Administrator, or

(B) the Administrator of the Federal Aviation Administration where there is alleged a failure of such Administrator to perform any act or duty under section 611 of the

Ante, p.
Federal Aviation Act of 1958 which is not discretionary with such Administrator.

Jurisdiction.
The district courts of the United States shall have jurisdiction, without regard to the amount in controversy, to restrain such person from violating such noise control requirement or to order such Administrator to perform such act or duty, as the case may be.

(b) No action may be commenced—

(1) under subsection (a)(1)—

Notice.
(A) prior to sixty days after the plaintiff has given notice of the violation (i) to the Administrator of the Environmental Protection Agency (and to the Federal Aviation Administrator in the case of a violation of a noise control requirement under such section 611) and (ii) to any alleged violator of such requirement, or

(B) if an Administrator has commenced and is diligently prosecuting a civil action to require compliance with the noise control requirement, but in any such action in a court of the United States any person may intervene as a matter of right, or

(2) under subsection (a)(2) prior to sixty days after the plaintiff has given notice to the defendant that he will commence such action.

Notice under this subsection shall be given in such manner as the Administrator of the Environmental Protection Agency shall prescribe by regulation.

Intervention.
(c) In an action under this section, the Administrator of the Environmental Protection Agency, if not a party, may intervene as a matter of right. In an action under this section respecting a noise control requirement under section 611 of the Federal Aviation Act of 1958,

the Administrator of the Federal Aviation Administration, if not a party, may also intervene as a matter of right.

(d) The court, in issuing any final order in any action brought pursuant to subsection (a) of this section, may award costs of litigation (including reasonable attorney and expert witness fees) to any party, whenever the court determines such an award is appropriate. Litigation costs.

(e) Nothing in this section shall restrict any right which any person (or class of persons) may have under any statute or common law to seek enforcement of any noise control requirement or to seek any other relief (including relief against an Administrator).

(f) For purposes of this section, the term "noise control requirement" means paragraph (1), (2), (3), (4), or (5) of section 10(a), or a standard, rule, or regulation issued under section 17 or 18 of this Act or under section 611 of the Federal Aviation Act of 1958. "Noise control requirement."

Ante, p.

RECORDS, REPORTS, AND INFORMATION

SEC. 13. (a) Each manufacturer of a product to which regulations under section 6 or section 8 apply shall—

(1) establish and maintain such records, make such reports, provide such information, and make such tests, as the Administrator may reasonably require to enable him to determine whether such manufacturer has acted or is acting in compliance with this Act,

(2) upon request of an officer or employee duly designated by the Administrator, permit such officer or employee at reasonable times to have access to such information and the results of such tests and to copy such records, and

(3) to the extent required by regulations of the Administrator, make products coming off the assembly line or otherwise in the hands of the manufacturer available for testing by the Administrator.

(b)(1) All information obtained by the Administrator or his representatives pursuant to subsection (a) of this section, which information contains or relates to a trade secret or other matter referred to in section 1905 of title 18 of the United States Code, shall be considered confidential for the purpose of that section, except that such information may be disclosed to other Federal officers or employees, in whose possession it shall remain confidential, or when relevant to the matter in controversy in any proceeding under this Act. Confidential information.

62 Stat. 791. Disclosure.

(2) Nothing in this subsection shall authorize the withholding of information by the Administrator, or by any officers or employees under his control, from the duly authorized committees of the Congress.

(c) Any person who knowingly makes any false statement, representation, or certification in any application, record, report, plan, or other document filed or required to be maintained under this Act or who falsifies, tampers with, or knowingly renders inaccurate any monitoring device or method required to be maintained under this Act, shall upon conviction be punished by a fine of not more than $10,000, or by imprisonment for not more than six months, or by both. Violations and penalties.

RESEARCH, TECHNICAL ASSISTANCE, AND PUBLIC INFORMATION

SEC. 14. In furtherance of his responsibilities under this Act and to complement, as necessary, the noise-research programs of other Federal agencies, the Administrator is authorized to:

(1) Conduct research, and finance research by contract with any person, on the effects, measurement, and control of noise, including but not limited to—

(A) investigation of the psychological and physiological effects of noise on humans and the effects of noise on domestic animals, wildlife, and property, and determination of acceptable levels of noise on the basis of such effects;

(B) development of improved methods and standards for measurement and monitoring of noise, in cooperation with the National Bureau of Standards, Department of Commerce; and

(C) determination of the most effective and practicable means of controlling noise emission.

(2) Provide technical assistance to State and local governments to facilitate their development and enforcement of ambient noise standards, including but not limited to—

(A) advice on training of noise-control personnel and on selection and operation of noise-abatement equipment; and

(B) preparation of model State or local legislation for noise control.

(3) Disseminate to the public information on the effects of noise, acceptable noise levels, and techniques for noise measurement and control.

DEVELOPMENT OF LOW-NOISE-EMISSION PRODUCTS

SEC. 15. (a) For the purpose of this section:

(1) The term "Committee" means the Low-Noise-Emission Product Advisory Committee.

(2) The term "Federal Government" includes the legislative, executive, and judicial branches of the Government of the United States, and the government of the District of Columbia.

(3) The term "low-noise-emission product" means any product which emits noise in amounts significantly below the levels specified in noise emission standards under regulations applicable under section 6 at the time of procurement to that type of product.

(4) The term "retail price" means (A) the maximum statutory price applicable to any type of product; or (B) in any case where there is no applicable maximum statutory price, the most recent procurement price paid for any type of product.

(b)(1) The Administrator shall determine which products qualify as low-noise-emission products in accordance with the provisions of this section.

(2) The Administrator shall certify any product—

(A) for which a certification application has been filed in accordance with paragraph (5)(A) of this subsection;

(B) which is a low-noise-emission product as determined by the Administrator; and

(C) which he determines is suitable for use as a substitute for a type of product at that time in use by agencies of the Federal Government.

(3) The Administrator may establish a Low-Noise-Emission Product Advisory Committee to assist him in determining which products qualify as low-noise-emission products for purposes of this section. The Committee shall include the Administrator or his designee, a representative of the National Bureau of Standards, and representatives of such other Federal agencies and private individuals as the Administrator may deem necessary from time to time. Any member of the Committee not employed on a full-time basis by the United States may receive the daily equivalent of the annual rate of basic pay in effect for grade GS–18 of the General Schedule for each day such

110

member is engaged upon work of the Committee. Each member of Travel expenses. the Committee shall be reimbursed for travel expenses, including per diem in lieu of subsistence as authorized by section 5703 of title 5, United States Code, for persons in the Government service employed 80 Stat. 499; intermittently. 83 Stat. 190.

(4) Certification under this section shall be effective for a period of one year from the date of issuance.

(5)(A) Any person seeking to have a class or model of product certified under this section shall file a certification application in accordance with regulations prescribed by the Administrator.

(B) The Administrator shall publish in the Federal Register a Publication in notice of each application received. Federal Regis-
ter.

(C) The Administrator shall make determinations for the purpose of this section in accordance with procedures prescribed by him by regulation.

(D) The Administrator shall conduct whatever investigation is necessary, including actual inspection of the product at a place designated in regulations prescribed under subparagraph (A).

(E) The Administrator shall receive and evaluate written comments and documents from interested persons in support of, or in opposition to, certification of the class or model of product under consideration.

(F) Within ninety days after the receipt of a properly filed certification application the Administrator shall determine whether such product is a low-noise-emission product for purposes of this section. If the Administrator determines that such product is a low-noise-emission product, then within one hundred and eighty days of such determination the Administrator shall reach a decision as to whether such product is a suitable substitute for any class or classes of products presently being purchased by the Federal Government for use by its agencies.

(G) Immediately upon making any determination or decision under Publication in subparagraph (F), the Administrator shall publish in the Federal Federal Regis- Register notice of such determination or decision, including reasons ter. therefor.

(c)(1) Certified low-noise-emission products shall be acquired by purchase or lease by the Federal Government for use by the Federal Government in lieu of other products if the Administrator of General Services determines that such certified products have procurement costs which are no more than 125 per centum of the retail price of the least expensive type of product for which they are certified substitutes.

(2) Data relied upon by the Administrator in determining that a product is a certified low-noise-emission product shall be incorporated in any contract for the procurement of such product.

(d) The procuring agency shall be required to purchase available certified low-noise-emission products which are eligible for purchase to the extent they are available before purchasing any other products for which any low-noise-emission product is a certified substitute. In making purchasing selections between competing eligible certified low-noise-emission products, the procuring agency shall give priority to any class or model which does not require extensive periodic maintenance to retain its low-noise-emission qualities or which does not involve operating costs significantly in excess of those products for which it is a certified substitute.

(e) For the purpose of procuring certified low-noise-emission Statutory price products any statutory price limitations shall be waived. limitations,

(f) The Administrator shall, from time to time as he deems appro- waiver. priate, test the emissions of noise from certified low-noise-emission products purchased by the Federal Government. If at any time he finds that the noise-emission levels exceed the levels on which certification under this section was based, the Administrator shall give the supplier of such product written notice of this finding, issue public notice of it, and give the supplier an opportunity to make necessary repairs, adjustments, or replacements. If no such repairs, adjustments, or replacements are made within a period to be set by the Administrator, he may order the supplier to show cause why the product involved should be eligible for recertification.

Appropriation.
(g) There are authorized to be appropriated for paying additional amounts for products pursuant to, and for carrying out the provisions of, this section, $1,000,000 for the fiscal year ending June 30, 1973, and $2,000,000 for each of the two succeeding fiscal years.

(h) The Administrator shall promulgate the procedures required to implement this section within one hundred and eighty days after the date of enactment of this Act.

JUDICIAL REVIEW; WITNESSES

SEC. 16. (a) A petition for review of action of the Administrator of the Environmental Protection Agency in promulgating any standard or regulation under section 6, 17, or 18 of this Act or any labeling regulation under section 8 of this Act may be filed only in the United States Court of Appeals for the District of Columbia Circuit, and a petition for review of action of the Administrator of the Federal Aviation Administration in promulgating any standard or regulation under
Ante, p. 1239.
section 611 of the Federal Aviation Act of 1958 may be filed only in such court. Any such petition shall be filed within ninety days from the date of such promulgation, or after such date if such petition is based solely on grounds arising after such ninetieth day. Action of either Administrator with respect to which review could have been obtained under this subsection shall not be subject to judicial review in civil or criminal proceedings for enforcement.

(b) If a party seeking review under this Act applies to the court for leave to adduce additional evidence, and shows to the satisfaction of the court that the information is material and was not available at the time of the proceeding before the Administrator of such Agency or Administration (as the case may be), the court may order such additional evidence (and evidence in rebuttal thereof) to be taken before such Administrator, and to be adduced upon the hearing, in such manner and upon such terms and conditions as the court may deem proper. Such Administrator may modify his findings as to the facts, or make new findings, by reason of the additional evidence so taken, and he shall file with the court such modified or new findings, and his recommendation, if any, for the modification or setting aside of his original order, with the return of such additional evidence.

(c) With respect to relief pending review of an action by either Administrator, no stay of an agency action may be granted unless the reviewing court determines that the party seeking such stay is (1) likely to prevail on the merits in the review proceeding and (2) will suffer irreparable harm pending such proceeding.

Subpenas.
(d) For the purpose of obtaining information to carry out this Act, the Administrator of the Environmental Protection Agency may issue subpenas for the attendance and testimony of witnesses and the production of relevant papers, books, and documents, and he may administer oaths. Witnesses summoned shall be paid the same fees and mileage that are paid witnesses in the courts of the United States. In cases of contumacy or refusal to obey a subpena served upon any person under this subsection, the district court of the United States for any district in which such person is found or resides or transacts business, upon application by the United States and after notice to such person,

112

shall have jurisdiction to issue an order requiring such person to appear and give testimony before the Administrator, to appear and produce papers, books, and documents before the Administrator, or both, and any failure to obey such order of the court may be punished by such court as a contempt thereof.

SEC. 17. (a) (1) Within nine months after the date of enactment of this Act, the Administrator shall publish proposed noise emission regulations for surface carriers engaged in interstate commerce by railroad. Such proposed regulations shall include noise emission standards setting such limits on noise emissions resulting from operation of the equipment and facilities of surface carriers engaged in interstate commerce by railroad which reflect the degree of noise reduction achievable through the application of the best available technology, taking into account the cost of compliance. These regulations shall be in addition to any regulations that may be proposed under section 6 of this Act. Regulations.

(2) Within ninety days after the publication of such regulations as may be proposed under paragraph (1) of this subsection, and subject to the provisions of section 16 of this Act, the Administrator shall promulgate final regulations. Such regulations may be revised, from time to time, in accordance with this subsection.

(3) Any standard or regulation, or revision thereof, proposed under this subsection shall be promulgated only after consultation with the Secretary of Transportation in order to assure appropriate consideration for safety and technological availability.

(4) Any regulation or revision thereof promulgated under this subsection shall take effect after such period as the Administrator finds necessary, after consultation with the Secretary of Transportation, to permit the development and application of the requisite technology, giving appropriate consideration to the cost of compliance within such period.

(b) The Secretary of Transportation, after consultation with the Administrator, shall promulgate regulations to insure compliance with all standards promulgated by the Administrator under this section. The Secretary of Transportation shall carry out such regulations through the use of his powers and duties of enforcement and inspection authorized by the Safety Appliance Acts, the Interstate Commerce Act, and the Department of Transportation Act. Regulations promulgated under this section shall be subject to the provisions of sections 10, 11, 12, and 16 of this Act. 27 Stat. 531. 45 USC 1. 24 Stat. 379. 49 USC prec. 1 note.

(c) (1) Subject to paragraph (2) but notwithstanding any other provision of this Act, after the effective date of a regulation under this section applicable to noise emissions resulting from the operation of any equipment or facility of a surface carrier engaged in interstate commerce by railroad, no State or political subdivision thereof may adopt or enforce any standard applicable to noise emissions resulting from the operation of the same equipment or facility of such carrier unless such standard is identical to a standard applicable to noise emissions resulting from such operation prescribed by any regulation under this section. 80 Stat. 931. 49 USC 1651 note.

(2) Nothing in this section shall diminish or enhance the rights of any State or political subdivision thereof to establish and enforce standards or controls on levels of environmental noise, or to control, license, regulate, or restrict the use, operation, or movement of any product if the Administrator, after consultation with the Secretary of Transportation, determines that such standard, control, license, regulation, or restriction is necessitated by special local conditions and is not in conflict with regulations promulgated under this section.

113

(d) The terms "carrier" and "railroad" as used in this section shall have the same meaning as such terms have under the first section of the Act of February 17, 1911 (45 U.S.C. 22).

MOTOR CARRIER NOISE EMISSION STANDARDS

Regulations.

SEC. 18. (a)(1) Within nine months after the date of enactment of this Act, the Administrator shall publish proposed noise emission regulations for motor carriers engaged in interstate commerce. Such proposed regulations shall include noise emission standards setting such limits on noise emissions resulting from operation of motor carriers engaged in interstate commerce which reflect the degree of noise reduction achievable through the application of the best available technology, taking into account the cost of compliance. These regulations shall be in addition to any regulations that may be proposed under section 6 of this Act.

(2) Within ninety days after the publication of such regulations as may be proposed under paragraph (1) of this subsection, and subject to the provisions of section 16 of this Act, the Administrator shall promulgate final regulations. Such regulations may be revised from time to time, in accordance with this subsection.

(3) Any standard or regulation, or revision thereof, proposed under this subsection shall be promulgated only after consultation with the Secretary of Transportation in order to assure appropriate consideration for safety and technological availability.

(4) Any regulation or revision thereof promulgated under this subsection shall take effect after such period as the Administrator finds necessary, after consultation with the Secretary of Transportation, to permit the development and application of the requisite technology, giving appropriate consideration to the cost of compliance within such period.

(b) The Secretary of Transportation, after consultation with the Administrator shall promulgate regulations to insure compliance with all standards promulgated by the Administrator under this section. The Secretary of Transportation shall carry out such regulations through the use of his powers and duties of enforcement and inspection authorized by the Interstate Commerce Act and the Department of Transportation Act. Regulations promulgated under this section shall be subject to the provisions of sections 10, 11, 12, and 16 of this Act.

24 Stat. 379.
49 USC prec.
1 note.
80 Stat. 931.
49 USC 1651
note.

(c)(1) Subject to paragraph (2) of this subsection but notwithstanding any other provision of this Act, after the effective date of a regulation under this section applicable to noise emissions resulting from the operation of any motor carrier engaged in interstate commerce, no State or political subdivision thereof may adopt or enforce any standard applicable to the same operation of such motor carrier, unless such standard is identical to a standard applicable to noise emissions resulting from such operation prescribed by any regulation under this section.

(2) Nothing in this section shall diminish or enhance the rights of any State or political subdivision thereof to establish and enforce standards or controls on levels of environmental noise, or to control, license, regulate, or restrict the use, operation, or movement of any product if the Administrator, after consultation with the Secretary of Transportation, determines that such standard, control, license, regulation, or restriction is necessitated by special local conditions and is not in conflict with regulations promulgated under this section.

(d) For purposes of this section, the term "motor carrier" includes a common carrier by motor vehicle, a contract carrier by motor vehicle,

114

and a private carrier of property by motor vehicle as those terms are defined by paragraphs (14), (15), and (17) of section 203(a) of the Interstate Commerce Act (49 U.S.C. 303(a)).

49 Stat. 545;
54 Stat. 920;
71 Stat. 411.

AUTHORIZATION OF APPROPRIATIONS

SEC. 19. There is authorized to be appropriated to carry out this Act (other than section 15) $3,000,000 for the fiscal year ending June 30, 1973; $6,000,000 for the fiscal year ending June 30, 1974; and $12,000,000 for the fiscal year ending June 30, 1975.

Approved October 27, 1972.

LEGISLATIVE HISTORY:

HOUSE REPORT No. 92-842 (Comm. on Interstate and Foreign Commerce).
SENATE REPORT No. 92-1160 accompanying S. 3342 (Comm. on Public Works).
CONGRESSIONAL RECORD, Vol. 118 (1972):
    Feb. 29, considered and passed House.
    Oct. 12, 13, considered and passed Senate, amended, in lieu of S. 3342.
    Oct. 18, House concurred in Senate amendment, with an amendment;
        Senate concurred in House amendment.
WEEKLY COMPILATION OF PRESIDENTIAL DOCUMENTS, Vol. 8, No. 44:
    Oct. 28, Presidential statement.

Summary

The Noise Control Act of 1972 was enacted to establish a means for effective coordination of Federal research and activities in noise control and to authorize the establishment of Federal noise emission standards for products distributed in commerce. In addition, the Act also provides information to the public respecting the noise emission and noise reduction characteristics of such products.

Review Questions

1. What are two specific purposes of the Noise Control Act of 1972?

2. Identify one economic factor that aided the passage of the Noise Control Act of 1972?

3. Identify one political factor that aided the passage of the Noise Control Act of 1972?

4. Identify one social factor that aided the passage of the Noise Control Act of 1972?

5. What effect has the Noise Control Act of 1972 had on the manufacture of new products?

6. What are the strengths of the Noise Control Act of 1972 with regard to current pollution control problems?

7. What are two weaknesses of the Noise Control Act of 1972 with regard to current pollution control problems?

clean water

OBJECTIVES:

The student should be able to identify the specific purposes of the Water Pollution Control Act.

The student should be able to describe those economic, political, and social factors leading to the passage of the Water Pollution Control Act.

The student should be able to describe the effect the Water Pollution Control Act has had on various types of industries.

The student should be able to identify the strengths and weaknesses of the Water Pollution Control Act as they relate to current pollution control problems and energy conservation measures.

INTRODUCTION:

O ur streams and rivers, our fresh water lakes, our salty bays and estuaries—these life-giving waters are among America's most precious natural endowments.

Yet today, many of our waters are grossly polluted by a staggering load of waste materials from farm, factory, and home, and there is scarcely a stream that does not bear some mark of man's abuse. The list of "most polluted" rivers spans the continent.

Certain pollutants, such as the phosphates, provide an excess of nutrients which disturb the ecological balance of our lakes and, by stimulating plant growth, greatly accelerate the otherwise slow, natural aging process. Lake Erie—not dead

TOWARD A NEW ENVIRONMENTAL ETHIC, U.S. Government Printing Office: 1971-O — 443-062, pp. 14-17.

but surely dying—is an outstanding example of this "eutrophication" process.

Contamination of our coastal waters prevents the harvesting of fish and shellfish in many areas. Dredging and filling operations threaten the estuarine waters that nurture aquatic life. Oil fouls our beaches and destroys fish and sea birds.

The ocean depths themselves are showing the effects; and far from the sources of pollution, polar bears and penguins carry DDT in their fatty tissues.

Even water far below the ground—the precious moisture that serves so many municipal drinking systems—faces pollution danger from poisonous wastes pumped into the earth.

The pollutants which clog America's waters are a mixed brew, and come from millions of sources:

• More than 1,300 communities still discharge their sewage into the waterways without any treatment whatever. An equal number employ only primary treatment, removing 30 to 40 percent of some pollutants. The waste flows from municipal systems are expected to increase by nearly four times over the next 50 years.

• Approximately 240,000 water-using industrial plants generate the largest volume and the most toxic of pollutants. The volume is growing.

• Oil spills from vessels and offshore drilling have produced tragic destruction along ocean beaches, while less spectacular spills, totaling thousands of barrels of oil, occur almost daily in waters across the nation.

Other important sources:

• Animal wastes from feedlots, fertilizer and pesticide runoff from fields and forests.

• Irrigation return-flows bearing fertilizer, pesticides and salts leached from the soil.

• Acid and sediment drainage from mining operations.

• Heated water, discharged principally by the electric power industry, which threatens aquatic life.

Federal clean water efforts were first launched in 1948 on a trial basis, and a permanent program established with passage of the Federal Water Pollution Control Act, in 1956. By the late 1960's a broad program based on federal/

118

state cooperation in establishing and enforcing water quality standards, river basin planning, and federal grants for construction of waste treatment facilities was in operation.

An important new mechanism for achieving clean water was provided when a Federal court decision in 1966 held that the Refuse Act of 1899 (previously applied only to debris that might obstruct navigation) outlaws industrial discharges of pollutants into navigable waters or their tributaries without a permit. Today, the provisions of this act are being increasingly invoked against polluters.

The Water Quality Improvement Act of 1970 included important new authorities to fight pollution, and today a massive effort has been launched to restore America's waters.

Major features of the current Federal Program are:

standards setting

• Under the Water Quality Act of 1965, each of the States, and the District of Columbia, Guam, Puerto Rico, and the Virgin Islands, have been in the process of establishing water quality standards for all interstate and coastal waters. These must be fully approved by EPA by January 1, 1972. The standards specify stream use classification (recreation, fish and wildlife propagation, public water supplies, industrial and agricultural uses); the quality of water required to support these uses; and complete plans for achieving and enforcing the desired levels of quality.

Over 90 percent of all interstate waters have been classified for either recreation or fish and wildlife propagation uses, which require high quality water.

enforcement

• States have first responsibility for enforcing the water quality standards. However, if the standards are violated, EPA may enforce them through procedures provided by the Water Quality Act or by initiating civil or criminal action under the Refuse Act.

One of the EPA Administrator's first official acts was to issue violation notices to three major cities placing them on notice that corrective action must be taken within 180 days to avoid prosecution; by the end of the 180-day period, agreement had been reached for joint federal-state-local construction of the needed treatment plants. In the Agency's first eight months, 47 such notices were issued to municipal or industrial polluters. These included 32 municipalities which discharge sewage into Lake Erie.

In several instances, EPA has initiated enforcement by convening a conference of federal and state officials, a procedure which is followed by a public hearing and court action against the violator if required. Of major importance are new conferences called to enforce standards covering the interstate waters of Long Island Sound in Connecticut and New York, of Galveston Bay and its tributaries in Texas, of western South Dakota, and of Pearl Harbor, Hawaii. Conferences concluded on Lake Michigan and Lake Superior resulted in major EPA actions to abate pollution.

As of mid-1971, more than 50 civil or criminal actions had been brought against industrial dischargers by the U.S. Department of Justice under the Refuse Act provisions. Fourteen civil suits had been concluded favorably by court-approved settlements. In civil actions against ten dischargers of mercury, a prompt reduction of total dischargers from 139 pounds to 2 pounds daily was brought about, with final disposition of the cases depending on EPA approval of plans for further reduction; in one case only, the discharging plant was shut down. Criminal prosecutions have generally resulted in convictions and assessment of fines; in one case, a violator was fined $125,000.

the refuse act permit program

By Presidential Order, a new permit program has been established under the provisions of the Refuse Act of 1899. Effective July 1, 1971, a permit from the U.S. Army Corps of Engineers is required for the discharge of any industrial waste into navigable waters or their tributaries. If, in the judgment of EPA, a discharge violates water quality standards, no permit can be issued. Violators are liable to swift Federal enforcement procedures.

construction of treatment facilities

EPA makes available federal funds to aid the construction of new or improved municipal sewage treatment plants. From the inception of the federal grant program in 1957 through July 1971, nearly $3 billion had been provided for over 12,000 projects. During the last half of 1971, the level of funding rose to $2 billion annually. By the end of 1974, EPA expects to have projects funded and underway that will provide secondary treatment for almost all municipal waste water.

oil spills

The Water Quality Improvement Act of 1970 prohibits the discharge of harmful quantities of oil into or upon the navigable waters of the United States or their shores. It applies to onshore and offshore facilities, as well as to vessels. The owner or operator is liable for cleanup costs and heavy penalties for knowingly discharging oil in harmful quantities.

EPA cooperates with the Coast Guard and other federal agencies in administering this Act and plays a primary role in implementing the National Contingency Plan for the removal of oil spills.

sewage from vessels

The 1970 legislation gave EPA authority to set performance standards, which will be enforced by the Coast Guard, for marine sanitation devices. Proposed standards were published in May 1971, requiring the equivalent of secondary treatment for vessel discharges. After all comments are considered, final standards are to be promulgated.

water hygiene

EPA establishes recommended health standards for municipal drinking water supplies, and waters used for recreation and shellfish-growing. These are generally used by States for assessing and improving the quality of water supplies within their boundaries. EPA may ban the use of unsafe water supplies on interstate carriers.

research and development

EPA supports research and demonstration projects, looking toward better means of controlling all forms of water pollution, with particular emphasis on finding improved ways to help municipalities and industry do the job.

other activities

EPA encourages effective river basin planning that takes into account all factors affecting water quality, provides expert technical assistance on difficult pollution problems, supports the training of much needed manpower in all aspects of pollution control, and gives financial and other assistance to States to help them strengthen their own control programs.

Public Law 92-500
92nd Congress, S. 2770
October 18, 1972

An Act

To amend the Federal Water Pollution Control Act.

Be it enacted by the Senate and House of Representatives of the United States of America in Congress assembled, That this Act may be cited as the "Federal Water Pollution Control Act Amendments of 1972".

Federal Water Pollution Control Act Amendments of 1972.
70 Stat. 498;
84 Stat. 91.
33 USC 1151 note.

Sec. 2. The Federal Water Pollution Control Act is amended to read as follows:

"TITLE I—RESEARCH AND RELATED PROGRAMS

"DECLARATION OF GOALS AND POLICY

"Sec. 101. (a) The objective of this Act is to restore and maintain the chemical, physical, and biological integrity of the Nation's waters. In order to achieve this objective it is hereby declared that, consistent with the provisions of this Act—

"(1) it is the national goal that the discharge of pollutants into the navigable waters be eliminated by 1985;

"(2) it is the national goal that wherever attainable, an interim goal of water quality which provides for the protection and propagation of fish, shellfish, and wildlife and provides for recreation in and on the water be achieved by July 1, 1983;

"(3) it is the national policy that the discharge of toxic pollutants in toxic amounts be prohibited;

"(4) it is the national policy that Federal financial assistance be provided to construct publicly owned waste treatment works;

"(5) it is the national policy that areawide waste treatment management planning processes be developed and implemented to assure adequate control of sources of pollutants in each State; and

"(6) it is the national policy that a major research and demonstration effort be made to develop technology necessary to eliminate the discharge of pollutants into the navigable waters, waters of the contiguous zone, and the oceans.

"(b) It is the policy of the Congress to recognize, preserve, and protect the primary responsibilities and rights of States to prevent, reduce, and eliminate pollution, to plan the development and use (including restoration, preservation, and enhancement) of land and water resources, and to consult with the Administrator in the exercise of his authority under this Act. It is further the policy of the Congress to support and aid research relating to the prevention, reduction, and elimination of pollution, and to provide Federal technical services and financial aid to State and interstate agencies and municipalities in connection with the prevention, reduction, and elimination of pollution.

"(c) It is further the policy of Congress that the President, acting through the Secretary of State and such national and international organizations as he determines appropriate, shall take such action as may be necessary to insure that to the fullest extent possible all foreign countries shall take meaningful action for the prevention, reduction, and elimination of pollution in their waters and in international waters and for the achievement of goals regarding the elimination of discharge of pollutants and the improvement of water quality to at least the same extent as the United States does under its laws.

"(d) Except as otherwise expressly provided in this Act, the Administrator of the Environmental Protection Agency (hereinafter in this Act called 'Administrator') shall administer this Act.

Administration.

FEDERAL WATER POLLUTION CONTROL ACT AMENDMENTS,
October 18, 1972, Public Law 92-500, 86 STAT. 816-904.

123

"(e) Public participation in the development, revision, and enforcement of any regulation, standard, effluent limitation, plan, or program established by the Administrator or any State under this Act shall be provided for, encouraged, and assisted by the Administrator and the States. The Administrator, in cooperation with the States, shall develop and publish regulations specifying minimum guidelines for public participation in such processes.

"(f) It is the national policy that to the maximum extent possible the procedures utilized for implementing this Act shall encourage the drastic minimization of paperwork and interagency decision procedures, and the best use of available manpower and funds, so as to prevent needless duplication and unnecessary delays at all levels of government.

"COMPREHENSIVE PROGRAMS FOR WATER POLLUTION CONTROL

"SEC. 102. (a) The Administrator shall, after careful investigation, and in cooperation with other Federal agencies, State water pollution control agencies, interstate agencies, and the municipalities and industries involved, prepare or develop comprehensive programs for preventing, reducing, or eliminating the pollution of the navigable waters and ground waters and improving the sanitary condition of surface and underground waters. In the development of such comprehensive programs due regard shall be given to the improvements which are necessary to conserve such waters for the protection and propagation of fish and aquatic life and wildlife, recreational purposes, and the withdrawal of such waters for public water supply, agricultural, industrial, and other purposes. For the purpose of this section, the Administrator is authorized to make joint investigations with any such agencies of the condition of any waters in any State or States, and of the discharges of any sewage, industrial wastes, or substance which may adversely affect such waters.

"(b) (1) In the survey or planning of any reservoir by the Corps of Engineers, Bureau of Reclamation, or other Federal agency, consideration shall be given to inclusion of storage for regulation of streamflow, except that any such storage and water releases shall not be provided as a substitute for adequate treatment or other methods of controlling waste at the source.

"(2) The need for and the value of storage for regulation of streamflow (other than for water quality) including but not limited to navigation, salt water intrusion, recreation, esthetics, and fish and wildlife, shall be determined by the Corps of Engineers, Bureau of Reclamation, or other Federal agencies.

"(3) The need for, the value of, and the impact of, storage for water quality control shall be determined by the Administrator, and his views on these matters shall be set forth in any report or presentation to Congress proposing authorization or construction of any reservoir including such storage.

"(4) The value of such storage shall be taken into account in determining the economic value of the entire project of which it is a part, and costs shall be allocated to the purpose of regulation of streamflow in a manner which will insure that all project purposes, share equitably in the benefits of multiple-purpose construction.

"(5) Costs of regulation of streamflow features incorporated in any Federal reservoir or other impoundment under the provisions of this Act shall be determined and the beneficiaries identified and if the benefits are widespread or national in scope, the costs of such features shall be nonreimbursable.

"(6) No license granted by the Federal Power Commission for a hydroelectric power project shall include storage for regulation of streamflow for the purpose of water quality control unless the Administrator shall recommend its inclusion and such reservoir storage capacity shall not exceed such proportion of the total storage required for the water quality control plan as the drainage area of such reservoir bears to the drainage area of the river basin or basins involved in such water quality control plan.

"(c) (1) The Administrator shall, at the request of the Governor of a State, or a majority of the Governors when more than one State is involved, make a grant to pay not to exceed 50 per centum of the administrative expenses of a planning agency for a period not to exceed three years, which period shall begin after the date of enactment of the Federal Water Pollution Control Act Amendments of 1972, if such agency provides for adequate representation of appropriate State, interstate, local, or (when appropriate) international interests in the basin or portion thereof involved and is capable of developing an effective, comprehensive water quality control plan for a basin or portion thereof.

"(2) Each planning agency receiving a grant under this subsection shall develop a comprehensive pollution control plan for the basin or portion thereof which—

"(A) is consistent with any applicable water quality standards, effluent and other limitations, and thermal discharge regulations established pursuant to current law within the basin;

"(B) recommends such treatment works as will provide the most effective and economical means of collection, storage, treatment, and elimination of pollutants and recommends means to encourage both municipal and industrial use of such works;

"(C) recommends maintenance and improvement of water quality within the basin or portion thereof and recommends methods of adequately financing those facilities as may be necessary to implement the plan; and

"(D) as appropriate, is developed in cooperation with, and is consistent with any comprehensive plan prepared by the Water Resources Council, any areawide waste management plans developed pursuant to section 208 of this Act, and any State plan developed pursuant to section 303(e) of this Act.

"(3) For the purposes of this subsection the term 'basin' includes, but is not limited to, rivers and their tributaries, streams, coastal waters, sounds, estuaries, bays, lakes, and portions thereof, as well as the lands drained thereby. "Basin."

"INTERSTATE COOPERATION AND UNIFORM LAWS

"SEC. 103. (a) The Administrator shall encourage cooperative activities by the States for the prevention, reduction, and elimination of pollution, encourage the enactment of improved and, so far as practicable, uniform State laws relating to the prevention, reduction, and elimination of pollution; and encourage compacts between States for the prevention and control of pollution.

"(b) The consent of the Congress is hereby given to two or more States to negotiate and enter into agreements or compacts, not in conflict with any law or treaty of the United States, for (1) cooperative effort and mutual assistance for the prevention and control of pollution and the enforcement of their respective laws relating thereto, and (2) the establishment of such agencies, joint or otherwise, as they may deem desirable for making effective such agreements and compacts. No such agreement or compact shall be binding or obligatory upon any State a party thereto unless and until it has been approved by the Congress.

"SEC. 104. (a) The Administrator shall establish national programs for the prevention, reduction, and elimination of pollution and as part of such programs shall—

"(1) in cooperation with other Federal, State, and local agencies, conduct and promote the coordination and acceleration of, research, investigations, experiments, training, demonstrations, surveys, and studies relating to the causes, effects, extent, prevention, reduction, and elimination of pollution;

"(2) encourage, cooperate with, and render technical services to pollution control agencies and other appropriate public or private agencies, institutions, and organizations, and individuals, including the general public, in the conduct of activities referred to in paragraph (1) of this subsection;

"(3) conduct, in cooperation with State water pollution control agencies and other interested agencies, organizations and persons, public investigations concerning the pollution of any navigable waters, and report on the results of such investigations;

"(4) establish advisory committees composed of recognized experts in various aspects of pollution and representatives of the public to assist in the examination and evaluation of research progress and proposals and to avoid duplication of research;

Water quality surveillance system, report.

"(5) in cooperation with the States, and their political subdivisions, and other Federal agencies establish, equip, and maintain a water quality surveillance system for the purpose of monitoring the quality of the navigable waters and ground waters and the contiguous zone and the oceans and the Administrator shall, to the extent practicable, conduct such surveillance by utilizing the resources of the National Aeronautics and Space Administration, the National Oceanic and Atmospheric Administration, the Geological Survey, and the Coast Guard, and shall report on such quality in the report required under subsection (a) of section 516; and

Report to Congress.

"(6) initiate and promote the coordination and acceleration of research designed to develop the most effective practicable tools and techniques for measuring the social and economic costs and benefits of activities which are subject to regulation under this Act; and shall transmit a report on the results of such research to the Congress not later than January 1, 1974.

"(b) In carrying out the provisions of subsection (a) of this section the Administrator is authorized to—

"(1) collect and make available, through publications and other appropriate means, the results of and other information, including appropriate recommendations by him in connection therewith, pertaining to such research and other activities referred to in paragraph (1) of subsection (a);

"(2) cooperate with other Federal departments and agencies, State water pollution control agencies, interstate agencies, other public and private agencies, institutions, organizations, industries involved, and individuals, in the preparation and conduct of such research and other activities referred to in paragraph (1) of subsection (a);

126

"(3) make grants to State water pollution control agencies, interstate agencies, other public or nonprofit private agencies. institutions, organizations, and individuals, for purposes stated in paragraph (1) of subsection (a) of this section;

"(4) contract with public or private agencies, institutions, organizations, and individuals, without regard to sections 3648 and 3709 of the Revised Statutes (31 U.S.C. 529; 41 U.S.C. 5), referred to in paragraph (1) of subsection (a);

"(5) establish and maintain research fellowships at public or nonprofit private educational institutions or research organizations;

"(6) collect and disseminate, in cooperation with other Federal departments and agencies, and with other public or private agencies, institutions, and organizations having related responsibilities, basic data on chemical, physical, and biological effects of varying water quality and other information pertaining to pollution and the prevention, reduction, and elimination thereof; and

"(7) develop effective and practical processes, methods, and prototype devices for the prevention, reduction, and elimination of pollution.

"(c) In carrying out the provisions of subsection (a) of this section the Administrator shall conduct research on, and survey the results of other scientific studies on, the harmful effects on the health or welfare of persons caused by pollutants. In order to avoid duplication of effort, the Administrator shall, to the extent practicable, conduct such research in cooperation with and through the facilities of the Secretary of Health, Education, and Welfare.

Pollutant effects, study.

HEW, cooperation.

"(d) In carrying out the provisions of this section the Administrator shall develop and demonstrate under varied conditions (including conducting such basic and applied research, studies, and experiments as may be necessary):

"(1) Practicable means of treating municipal sewage, and other waterborne wastes to implement the requirements of section 201 of this Act;

"(2) Improved methods and procedures to identify and measure the effects of pollutants, including those pollutants created by new technological developments; and

"(3) Methods and procedures for evaluating the effects on water quality of augmented streamflows to control pollution not susceptible to other means of prevention, reduction, or elimination.

"(e) The Administrator shall establish, equip, and maintain field laboratory and research facilities, including, but not limited to, one to be located in the northeastern area of the United States, one in the Middle Atlantic area, one in the southeastern area, one in the midwestern area, one in the southwestern area, one in the Pacific Northwest, and one in the State of Alaska, for the conduct of research, investigations, experiments, field demonstrations and studies, and training relating to the prevention, reduction and elimination of pollution. Insofar as practicable, each such facility shall be located near institutions of higher learning in which graduate training in such research might be carried out. In conjunction with the development of criteria under section 403 of this Act, the Administrator shall construct the facilities authorized for the National Marine Water Quality Laboratory established under this subsection.

Field research laboratories.

"(f) The Administrator shall conduct research and technical development work, and make studies, with respect to the quality of the waters of the Great Lakes, including an analysis of the present and

Great Lakes, water quality research.

127

projected future water quality of the Great Lakes under varying conditions of waste treatment and disposal, an evaluation of the water quality needs of those to be served by such waters, an evaluation of municipal, industrial, and vessel waste treatment and disposal practices with respect to such waters, and a study of alternate means of solving pollution problems (including additional waste treatment measures) with respect to such waters.

Treatment works, pilot training programs.

"(g)(1) For the purpose of providing an adequate supply of trained personnel to operate and maintain existing and future treatment works and related activities, and for the purpose of enhancing substantially the proficiency of those engaged in such activities, the Administrator shall finance pilot programs, in cooperation with State and interstate agencies, municipalities, educational institutions, and other organizations and individuals, of manpower development and training and retraining of persons in, on entering into, the field of operation and maintenance of treatment works and related activities. Such program and any funds expended for such a program shall supplement, not supplant, other manpower and training programs and funds available for the purposes of this paragraph. The Administrator is authorized, under such terms and conditions as he deems appropriate, to enter into agreements with one or more States, acting jointly or severally, or with other public or private agencies or institutions for the development and implementation of such a program.

Employment needs, forecasting.

"(2) The Administrator is authorized to enter into agreements with public and private agencies and institutions, and individuals to develop and maintain an effective system for forecasting the supply of, and demand for, various professional and other occupational categories needed for the prevention, reduction, and elimination of pollution in each region, State, or area of the United States and, from time to time, to publish the results of such forecasts.

"(3) In furtherance of the purposes of this Act, the Administrator is authorized to—

"(A) make grants to public or private agencies and institutions and to individuals for training projects, and provide for the conduct of training by contract with public or private agencies and institutions and with individuals without regard to sections 3648

31 USC 529. 41 USC 5.

and 3709 of the Revised Statutes;

"(B) establish and maintain research fellowships in the Environmental Protection Agency with such stipends and allowances, including traveling and subsistence expenses, as he may deem necessary to procure the assistance of the most promising research fellows; and

"(C) provide, in addition to the program established under paragraph (1) of this subsection, training in technical matters relating to the causes, prevention, reduction, and elimination of pollution for personnel of public agencies and other persons with suitable qualifications.

Report to President, transmittal to Congress.

"(4) The Administrator shall submit, through the President, a report to the Congress not later than December 31, 1973, summarizing the actions taken under this subsection and the effectiveness of such actions, and setting forth the number of persons trained, the occupational categories for which training was provided, the effectiveness of other Federal, State, and local training programs in this field, together with estimates of future needs, recommendations on improving training programs, and such other information and recommendations, including legislative recommendations, as he deems appropriate.

Lake pollution.

"(h) The Administrator is authorized to enter into contracts with, or make grants to, public or private agencies and organizations and

128

individuals for (A) the purpose of developing and demonstrating new or improved methods for the prevention, removal, reduction, and elimination of pollution in lakes, including the undesirable effects of nutrients and vegetation, and (B) the construction of publicly owned research facilities for such purpose.

"(i) The Administrator, in cooperation with the Secretary of the department in which the Coast Guard is operating, shall— Oil pollution control, studies.

"(1) engage in such research, studies, experiments, and demonstrations as he deems appropriate, relative to the removal of oil from any waters and to the prevention, control, and elimination of oil and hazardous substances pollution;

"(2) publish from time to time the results of such activities; and

"(3) from time to time, develop and publish in the Federal Register specifications and other technical information on the various chemical compounds used in the control of oil and hazardous substances spills. Publication in Federal Register.

In carrying out this subsection, the Administrator may enter into contracts with, or make grants to, public or private agencies and organizations and individuals.

"(j) The Secretary of the department in which the Coast Guard is operating shall engage in such research, studies, experiments, and demonstrations as he deems appropriate relative to equipment which is to be installed on board a vessel and is designed to receive, retain, treat, or discharge human body wastes and the wastes from toilets and other receptacles intended to receive or retain body wastes with particular emphasis on equipment to be installed on small recreational vessels. The Secretary of the department in which the Coast Guard is operating shall report to Congress the results of such research, studies, experiments, and demonstrations prior to the effective date of any regulations established under section 312 of this Act. In carrying out this subsection the Secretary of the department in which the Coast Guard is operating may enter into contracts with, or make grants to, public or private organizations and individuals. Vessels, solid waste disposal equipment.

Report to Congress.

"(k) In carrying out the provisions of this section relating to the conduct by the Administrator of demonstration projects and the development of field laboratories and research facilities, the Administrator may acquire land and interests therein by purchase, with appropriated or donated funds, by donation, or by exchange for acquired or public lands under his jurisdiction which he classifies as suitable for disposition. The values of the properties so exchanged either shall be approximately equal, or if they are not approximately equal, the values shall be equalized by the payment of cash to the grantor or to the Administrator as the circumstances require. Land acquisition.

"(l)(1) The Administrator shall, after consultation with appropriate local, State, and Federal agencies, public and private organizations, and interested individuals, as soon as practicable but not later than January 1, 1973, develop and issue to the States for the purpose of carrying out this Act the latest scientific knowledge available in indicating the kind and extent of effects on health and welfare which may be expected from the presence of pesticides in the water in varying quantities. He shall revise and add to such information whenever necessary to reflect developing scientific knowledge. Pesticides, effects and control.

"(2) The President shall, in consultation with appropriate local, State, and Federal agencies, public and private organizations, and interested individuals, conduct studies and investigations of methods to control the release of pesticides into the environment which study shall include examination of the persistency of pesticides in the water environment and alternatives thereto. The President shall submit reports, from time to time, on such investigations to Congress together with his recommendations for any necessary legislation.

"(m)(1) The Administrator shall, in an effort to prevent degradation of the environment from the disposal of waste oil, conduct a study of (A) the generation of used engine, machine, cooling, and similar waste oil, including quantities generated, the nature and quality of such oil, present collecting methods and disposal practices, and alternate uses of such oil; (B) the long-term, chronic biological effects of the disposal of such waste oil; and (C) the potential market for such oils, including the economic and legal factors relating to the sale of products made from such oils, the level of subsidy, if any, needed to encourage the purchase by public and private nonprofit agencies of products from such oil, and the practicability of Federal procurement, on a priority basis, of products made from such oil. In conducting such study, the Administrator shall consult with affected industries and other persons.

"(2) The Administrator shall report the preliminary results of such study to Congress within six months after the date of enactment of the Federal Water Pollution Control Act Amendments of 1972, and shall submit a final report to Congress within 18 months after such date of enactment.

"(n)(1) The Administrator shall, in cooperation with the Secretary of the Army, the Secretary of Agriculture, the Water Resources Council, and with other appropriate Federal, State, interstate, or local public bodies and private organizations, institutions, and individuals, conduct and promote, and encourage contributions to, continuing comprehensive studies of the effects of pollution, including sedimentation, in the estuaries and estuarine zones of the United States on fish and wildlife, on sport and commercial fishing, on recreation, on water supply and water power, and on other beneficial purposes. Such studies shall also consider the effect of demographic trends, the exploitation of mineral resources and fossil fuels, land and industrial development, navigation, flood and erosion control, and other uses of estuaries and estuarine zones upon the pollution of the waters therein.

"(2) In conducting such studies, the Administrator shall assemble, coordinate, and organize all existing pertinent information on the Nation's estuaries and estuarine zones; carry out a program of investigations and surveys to supplement existing information in representative estuaries and estuarine zones; and identify the problems and areas where further research and study are required.

"(3) The Administrator shall submit to Congress, from time to time, reports of the studies authorized by this subsection but at least one such report during any three year period. Copies of each such report shall be made available to all interested parties, public and private.

"(4) For the purpose of this subsection, the term 'estuarine zones' means an environmental system consisting of an estuary and those transitional areas which are consistently influenced or affected by water from an estuary such as, but not limited to, salt marshes, coastal and intertidal areas, bays, harbors, lagoons, inshore waters, and channels,

and the term 'estuary' means all or part of the mouth of a river or stream or other body of water having unimpaired natural connection with open sea and within which the sea water is measurably diluted with fresh water derived from land drainage.

"(o)(1) The Administrator shall conduct research and investigations on devices, systems, incentives, pricing policy, and other methods of reducing the total flow of sewage, including, but not limited to, unnecessary water consumption in order to reduce the requirements for, and the costs of, sewage and waste treatment services. Such research and investigations shall be directed to develop devices, systems, policies, and methods capable of achieving the maximum reduction of unnecessary water consumption.

"(2) The Administrator shall report the preliminary results of such studies and investigations to the Congress within one year after the date of enactment of the Federal Water Pollution Control Act Amendments of 1972, and annually thereafter in the report required under subsection (a) of section 516. Such report shall include recommendations for any legislation that may be required to provide for the adoption and use of devices, systems, policies, or other methods of reducing water consumption and reducing the total flow of sewage. Such report shall include an estimate of the benefits to be derived from adoption and use of such devices, systems, policies, or other methods and also shall reflect estimates of any increase in private, public, or other cost that would be occasioned thereby.

"(p) In carrying out the provisions of subsection (a) of this section the Administrator shall, in cooperation with the Secretary of Agriculture, other Federal agencies, and the States, carry out a comprehensive study and research program to determine new and improved methods and the better application of existing methods of preventing, reducing, and eliminating pollution from agriculture, including the legal, economic, and other implications of the use of such methods.

"(q) (1) The Administrator shall conduct a comprehensive program of research and investigation and pilot project implementation into new and improved methods of preventing, reducing, storing, collecting, treating, or otherwise eliminating pollution from sewage in rural and other areas where collection of sewage in conventional, community-wide sewage collection systems is impractical, uneconomical, or otherwise infeasible, or where soil conditions or other factors preclude the use of septic tank and drainage field systems.

"(2) The Administrator shall conduct a comprehensive program of research and investigation and pilot project implementation into new and improved methods for the collection and treatment of sewage and other liquid wastes combined with the treatment and disposal of solid wastes.

"(r) The Administrator is authorized to make grants to colleges and universities to conduct basic research into the structure and function of fresh water aquatic ecosystems, and to improve understanding of the ecological characteristics necessary to the maintenance of the chemical, physical, and biological integrity of freshwater aquatic ecosystems.

"(s) The Administrator is authorized to make grants to one or more institutions of higher education (regionally located and to be designated as 'River Study Centers') for the purpose of conducting and reporting on interdisciplinary studies on the nature of river systems, including hydrology, biology, ecology, economics, the relationship between river uses and land uses, and the effects of development within river basins on river systems and on the value of water resources and water related activities. No such grant in any fiscal year shall exceed $1,000,000.

"(t) The Administrator shall, in cooperation with State and Federal agencies and public and private organizations, conduct continuing comprehensive studies of the effects and methods of control of thermal discharges. In evaluating alternative methods of control the studies shall consider (1) such data as are available on the latest available technology, economic feasibility including cost-effectiveness analysis, and (2) the total impact on the environment, considering not only water quality but also air quality, land use, and effective utilization and conservation of fresh water and other natural resources. Such studies shall consider methods of minimizing adverse effects and maximizing beneficial effects of thermal discharges.

Public information. The results of these studies shall be reported by the Administrator as soon as practicable, but not later than 270 days after enactment of this subsection, and shall be made available to the public and the States, and considered as they become available by the Administrator in carrying out section 316 of this Act and by the States in proposing thermal water quality standards.

Appropriations. "(u) There is authorized to be appropriated (1) $100,000,000 per fiscal year for the fiscal year ending June 30, 1973, and the fiscal year ending June 30, 1974, for carrying out the provisions of this section other than subsections (g) (1) and (2), (p), (r), and (t); (2) not to exceed $7,500,000 for fiscal year 1973 for carrying out the provisions of subsection (g) (1); (3) not to exceed $2.500,000 for fiscal year 1973 for carrying out the provisions of subsection (g) (2); (4) not to exceed $10,000,000 for each of the fiscal years ending June 30, 1973, and June 30. 1974, for carrying out the provisions of subsection (p); (5) not to exceed $15,000,000 per fiscal year for the fiscal years ending June 30, 1973, and June 30, 1974, for carrying out the provisions of subsection (r); and (6) not to exceed $10,000,000 per fiscal year for the fiscal years ending June 30, 1973, and June 30, 1974, for carrying out the provisions of subsection (t).

"GRANTS FOR RESEARCH AND DEVELOPMENT

Environmental Protection Agency, demonstration projects. "SEC. 105. (a) The Administrator is authorized to conduct in the Environmental Protection Agency, and to make grants to any State, municipality, or intermunicipal or interstate agency for the purpose of assisting in the development of—

"(1) any project which will demonstrate a new or improved method of preventing, reducing, and eliminating the discharge into any waters of pollutants from sewers which carry storm water or both storm water and pollutants; or

"(2) any project which will demonstrate advanced waste treatment and water purification methods (including the temporary use of new or improved chemical additives which provide substantial immediate improvement to existing treatment processes), or new or improved methods of joint treatment systems for municipal and industrial wastes;

and to include in such grants such amounts as are necessary for the purpose of reports, plans, and specifications in connection therewith.

"(b) The Administrator is authorized to make grants to any State or States or interstate agency to demonstrate, in river basins or portions thereof. advanced treatment and environmental enhancement techniques to control pollution from all sources, within such basins or portions thereof. including nonpoint sources, together with in stream water quality improvement techniques.

"(c) In order to carry out the purposes of section 301 of this Act, the Administrator is authorized to (1) conduct in the Environmental Protection Agency, (2) make grants to persons, and (3) enter into contracts with persons, for research and demonstration projects for prevention of pollution of any waters by industry including, but not limited to, the prevention, reduction. and elimination of the discharge of pollutants. No grant shall be made for any project under this subsection unless the Administrator determines that such project will develop or demonstrate a new or improved method of treating industrial wastes or otherwise prevent pollution by industry, which method shall have industrywide application.

"(d) In carrying out the provisions of this section, the Administrator shall conduct, on a priority basis, an accelerated effort to develop, refine, and achieve practical application of:

"(1) waste management methods applicable to point and nonpoint sources of pollutants to eliminate the discharge of pollutants, including, but not limited to, elimination of runoff of pollutants and the effects of pollutants from inplace or accumulated sources;

"(2) advanced waste treatment methods applicable to point and nonpoint sources, including inplace or accumulated sources of pollutants, and methods for reclaiming and recycling water and confining pollutants so they will not migrate to cause water or other environmental pollution; and

"(3) improved methods and procedures to identify and measure the effects of pollutants on the chemical, physical, and biological integrity of water, including those pollutants created by new technological developments.

"(e)(1) The Administrator is authorized to (A) make, in consultation with the Secretary of Agriculture, grants to persons for research and demonstration projects with respect to new and improved methods of preventing, reducing, and eliminating pollution from agriculture, and (B) disseminate, in cooperation with the Secretary of Agriculture, such information obtained under this subsection, section 104(p), and section 304 as will encourage and enable the adoption of such methods in the agricultural industry.

"(2) The Administrator is authorized, (A) in consultation with other interested Federal agencies, to make grants for demonstration projects with respect to new and improved methods of preventing, reducing, storing, collecting, treating, or otherwise eliminating pollution from sewage in rural and other areas where collection of sewage in conventional, community-wide sewage collection systems is impractical, uneconomical, or otherwise infeasible, or where soil conditions or other factors preclude the use of septic tank and drainage field systems, and (B) in cooperation with other interested Federal and State agencies, to disseminate such information obtained under this subsection as will encourage and enable the adoption of new and improved methods developed pursuant to this subsection.

"(f) Federal grants under subsection (a) of this section shall be subject to the following limitations: Limitations.

"(1) No grant shall be made for any project unless such project shall have been approved by the appropriate State water pollution control agency or agencies and by the Administrator;

"(2) No grant shall be made for any project in an amount exceeding 75 per centum of cost thereof as determined by the Administrator; and

"(3) No grant shall be made for any project unless the Administrator determines that such project will serve as a useful demonstration for the purpose set forth in clause (1) or (2) of subsection (a).

"(g) Federal grants under subsections (c) and (d) of this section shall not exceed 75 per centum of the cost of the project.

"(h) For the purpose of this section there is authorized to be appropriated $75,000,000 per fiscal year for the fiscal year ending June 30, 1973, and the fiscal year ending June 30, 1974, and from such appropriations at least 10 per centum of the funds actually appropriated in each fiscal year shall be available only for the purposes of subsection (e). Appropriation.

133

"GRANTS FOR POLLUTION CONTROL PROGRAMS

State programs,
appropriations. "SEC. 106. (a) There are hereby authorized to be appropriated the following sums, to remain available until expended, to carry out the purposes of this section—

"(1) $60,000,000 for the fiscal year ending June 30, 1973; and
"(2) $75,000,000 for the fiscal year ending June 30, 1974;

for grants to States and to interstate agencies to assist them in administering programs for the prevention, reduction, and elimination of pollution, including enforcement directly or through appropriate State law enforcement officers or agencies.

Allotments. "(b) From the sums appropriated in any fiscal year, the Administrator shall make allotments to the several States and interstate agencies in accordance with regulations promulgated by him on the basis of the extent of the pollution problem in the respective States.

"(c) The Administrator is authorized to pay to each State and interstate agency each fiscal year either—

"(1) the allotment of such State or agency for such fiscal year under subsection (b), or

"(2) the reasonable costs as determined by the Administrator of developing and carrying out a pollution program by such State or agency during such fiscal year,

which ever amount is the lesser.

Limitations. "(d) No grant shall be made under this section to any State or interstate agency for any fiscal year when the expenditure of non-Federal funds by such State or interstate agency during such fiscal year for the recurrent expenses of carrying out its pollution control program are less than the expenditure by such State or interstate agency of non-Federal funds for such recurrent program expenses during the fiscal year ending June 30, 1971.

"(e) Beginning in fiscal year 1974 the Administrator shall not make any grant under this section to any State which has not provided or is not carrying out as a part of its program—

"(1) the establishment and operation of appropriate devices, methods, systems, and procedures necessary to monitor, and to compile and analyze data on (including classification according to eutrophic condition), the quality of navigable waters and to the extent practicable, ground waters including biological monitoring; and provision for annually updating such data and including it in the report required under section 305 of this Act;

"(2) authority comparable to that in section 504 of this Act and adequate contingency plans to implement such authority.

Conditions. "(f) Grants shall be made under this section on condition that—

"(1) Such State (or interstate agency) files with the Administrator within one hundred and twenty days after the date of enactment of this section:

"(A) a summary report of the current status of the State pollution control program, including the criteria used by the State in determining priority of treatment works; and

"(B) such additional information, data, and reports as the Administrator may require.

"(2) No federally assumed enforcement as defined in section 309 (a) (2) is in effect with respect to such State or interstate agency.

"(3) Such State (or interstate agency) submits within one hundred and twenty days after the date of enactment of this section and before July 1 of each year thereafter for the Administrator's approval its program for the prevention, reduction, and elimination of pollution in accordance with purposes and provisions of this Act in such form and content as the Administrator may prescribe.

134

"(g) Any sums allotted under subsection (b) in any fiscal year which are not paid shall be reallotted by the Administrator in accordance with regulations promulgated by him.

"MINE WATER POLLUTION CONTROL DEMONSTRATIONS

"SEC. 107. (a) The Administrator in cooperation with the Appalachian Regional Commission and other Federal agencies is authorized to conduct, to make grants for, or to contract for, projects to demonstrate comprehensive approaches to the elimination or control of acid or other mine water pollution resulting from active or abandoned mining operations and other environmental pollution affecting water quality within all or part of a watershed or river basin, including siltation from surface mining. Such projects shall demonstrate the engineering and economic feasibility and practicality of various abatement techniques which will contribute substantially to effective and practical methods of acid or other mine water pollution elimination or control, and other pollution affecting water quality, including techniques that demonstrate the engineering and economic feasibility and practicality of using sewage sludge materials and other municipal wastes to diminish or prevent pollution affecting water quality from acid, sedimentation, or other pollutants and in such projects to restore affected lands to usefulness for forestry, agriculture, recreation, or other beneficial purposes.

"(b) Prior to undertaking any demonstration project under this section in the Appalachian region (as defined in section 403 of the Appalachian Regional Development Act of 1965, as amended), the Appalachian Regional Commission shall determine that such demonstration project is consistent with the objectives of the Appalachian Regional Development Act of 1965, as amended.

79 Stat. 21;
81 Stat. 266;
83 Stat. 215.
40 USC app. 403.
79 Stat. 5;
85 Stat. 173.
40 USC app. 1.

"(c) The Administrator, in selecting watersheds for the purposes of this section, shall be satisfied that the project area will not be affected adversely by the influx of acid or other mine water pollution from nearby sources.

"(d) Federal participation in such projects shall be subject to the conditions—

Federal participation, conditions.

"(1) that the State shall acquire any land or interests therein necessary for such project; and

"(2) that the State shall provide legal and practical protection to the project area to insure against any activities which will cause future acid or other mine water pollution.

"(e) There is authorized to be appropriated $30,000,000 to carry out the provisions of this section, which sum shall be available until expended.

Appropriation.

"POLLUTION CONTROL IN GREAT LAKES

"SEC. 108. (a) The Administrator, in cooperation with other Federal departments, agencies, and instrumentalities is authorized to enter into agreements with any State, political subdivision, interstate agency, or other public agency, or combination thereof, to carry out one or more projects to demonstrate new methods and techniques and to develop preliminary plans for the elimination or control of pollution, within all or any part of the watersheds of the Great Lakes. Such projects shall demonstrate the engineering and economic feasibility and practicality of removal of pollutants and prevention of any polluting matter from entering into the Great Lakes in the future and other reduction and remedial techniques which will contribute substantially to effective and practical methods of pollution prevention, reduction, or elimination.

Federal-State cooperation.

"(b) Federal participation in such projects shall be subject to the condition that the State, political subdivision, interstate agency, or other public agency, or combination thereof, shall pay not less than 25 per centum of the actual project costs, which payment may be in any form, including, but not limited to, land or interests therein that is needed for the project, and personal property or services the value of which shall be determined by the Administrator.

Appropriation.

"(c) There is authorized to be appropriated $20,000,000 to carry out the provisions of subsections (a) and (b) of this section, which sum shall be available until expended.

Lake Erie
demonstration
program.

"(d)(1) In recognition of the serious conditions which exist in Lake Erie, the Secretary of the Army, acting through the Chief of Engineers, is directed to design and develop a demonstration waste water management program for the rehabilitation and environmental repair of Lake Erie. Prior to the initiation of detailed engineering and design, the program, along with the specific recommendations of the Chief of Engineers, and recommendations for its financing, shall be submitted to the Congress for statutory approval. This authority is in addition to, and not in lieu of, other waste water studies aimed at eliminating pollution emanating from select sources around Lake Erie.

Alternative
systems.

"(2) This program is to be developed in cooperation with the Environmental Protection Agency, other interested departments, agencies, and instrumentalities of the Federal Government, and the States and their political subdivisions. This program shall set forth alternative systems for managing waste water on a regional basis and shall provide local and State governments with a range of choice as to the type of system to be used for the treatment of waste water. These alternative systems shall include both advanced waste treatment technology and land disposal systems including aerated treatment-spray irrigation technology and will also include provisions for the disposal of solid wastes, including sludge. Such program should include measures to control point sources of pollution, area sources of pollution, including acid-mine drainage, urban runoff and rural runoff, and in place sources of pollution, including bottom loads, sludge banks, and polluted harbor dredgings.

Appropriation.

"(e) There is authorized to be appropriated $5,000,000 to carry out the provisions of subsection (d) of this section, which sum shall be available until expended.

"TRAINING GRANTS AND CONTRACTS

"SEC. 109. (a) The Administrator is authorized to make grants to or contracts with institutions of higher education, or combinations of such institutions, to assist them in planning, developing, strengthening, improving, or carrying out programs or projects for the preparation of undergraduate students to enter an occupation which involves the design, operation, and maintenance of treatment works, and other facilities whose purpose is water quality control. Such grants or contracts may include payment of all or part of the cost of programs or projects such as—

"(A) planning for the development or expansion of programs or projects for training persons in the operation and maintenance of treatment works;

"(B) training and retraining of faculty members;

"(C) conduct of short-term or regular session institutes for study by persons engaged in, or preparing to engage in, the preparation of students preparing to enter an occupation involving the operation and maintenance of treatment works;

"(D) carrying out innovative and experimental programs of cooperative education involving alternate periods of full-time or part-time academic study at the institution and periods of full-time or part-time employment involving the operation and maintenance of treatment works; and

"(E) research into, and development of, methods of training students or faculty, including the preparation of teaching materials and the planning of curriculum.

"(b)(1) The Administrator may pay 100 per centum of any additional cost of construction of a treatment works required for a facility to train and upgrade waste treatment works operation and maintenance personnel.

"(2) The Administrator shall make no more than one grant for such additional construction in any State (to serve a group of States, where, in his judgment, efficient training programs require multi-State programs), and shall make such grant after consultation with and approval by the State or States on the basis of (A) the suitability of such facility for training operation and maintenance personnel for treatment works throughout such State or States; and (B) a commitment by the State agency or agencies to carry out at such facility a program of training approved by the Administrator.

"(3) The Administrator may make such grant out of the sums allocated to a State under section 205 of this Act, except that in no event shall the Federal cost of any such training facilities exceed $250,000.

Limitation.

"APPLICATION FOR TRAINING GRANT OR CONTRACT; ALLOCATION OF GRANTS OR CONTRACTS

"SEC. 110. (1) A grant or contract authorized by section 109 may be made only upon application to the Administrator at such time or times and containing such information as he may prescribe, except that no such application shall be approved unless it—

"(A) sets forth programs, activities, research, or development for which a grant is authorized under section 109 and describes the relation to any program set forth by the applicant in an application, if any, submitted pursuant to section 111;

"(B) provides such fiscal control and fund accounting procedures as may be necessary to assure proper disbursement of and accounting for Federal funds paid to the applicant under this section; and

"(C) provides for making such reports, in such form and containing such information, as the Administrator may require to carry out his functions under this section, and for keeping such records and for affording such access thereto as the Administrator may find necessary to assure the correctness and verification of such reports.

"(2) The Administrator shall allocate grants or contracts under section 109 in such manner as will most nearly provide an equitable distribution of the grants or contracts throughout the United States among institutions of higher education which show promise of being able to use funds effectively for the purpose of this section.

"(3)(A) Payment under this section may be used in accordance with regulations of the Administrator, and subject to the terms and conditions set forth in an application approved under paragraph (1), to pay part of the compensation of students employed in connection with the operation and maintenance of treatment works, other than as an employee in connection with the operation and maintenance of treatment works or as an employee in any branch of the Government of the United States, as part of a program for which a grant has been approved pursuant to this section.

"(B) Departments and agencies of the United States are encouraged, to the extent consistent with efficient administration, to enter into arrangements with institutions of higher education for the full-time, part-time, or temporary employment, whether in the competitive or excepted service, of students enrolled in programs set forth in applications approved under paragraph (1).

"AWARD OF SCHOLARSHIPS

"SEC. 111. (1) The Administrator is authorized to award scholarships in accordance with the provisions of this section for undergraduate study by persons who plan to enter an occupation involving the operation and maintenance of treatment works. Such scholarships shall be awarded for such periods as the Administrator may determine but not to exceed four academic years.

"(2) The Administrator shall allocate scholarships under this section among institutions of higher education with programs approved under the provisions of this section for the use of individuals accepted into such programs, in such manner and according to such plan as will insofar as practicable—

"(A) provide an equitable distribution of such scholarships throughout the United States; and

"(B) attract recent graduates of secondary schools to enter an occupation involving the operation and maintenance of treatment works.

"(3) The Administrator shall approve a program of any institution of higher education for the purposes of this section only upon application by the institution and only upon his finding—

"(A) that such program has a principal objective the education and training of persons in the operation and maintenance of treatment works;

"(B) that such program is in effect and of high quality, or can be readily put into effect and may reasonably be expected to be of high quality;

"(C) that the application describes the relation of such program to any program, activity, research, or development set forth by the applicant in an application, if any, submitted pursuant to section 110 of this Act; and

"(D) that the application contains satisfactory assurances that (i) the institution will recommend to the Administrator for the award of scholarships under this section, for study in such program, only persons who have demonstrated to the satisfaction of the institution a serious intent, upon completing the program, to enter an occupation involving the operation and maintenance of treatment works, and (ii) the institution will make reasonable continuing efforts to encourage recipients of scholarships under this section, enrolled in such program, to enter occupations involving the operation and maintenance of treatment works upon completing the program.

"(4)(A) The Administrator shall pay to persons awarded scholarships under this section such stipends (including such allowances for subsistence and other expenses for such persons and their dependents) as he may determine to be consistent with prevailing practices under comparable federally supported programs.

"(B) The Administrator shall (in addition to the stipends paid to persons under paragraph (1)) pay to the institution of higher education at which such person is pursuing his course of study such amount as he may determine to be consistent with prevailing practices under comparable federally supported programs.

"(5) A person awarded a scholarship under the provisions of this section shall continue to receive the payments provided in this section only during such periods as the Administrator finds that he is maintaining satisfactory proficiency and devoting full time to study or research in the field in which such scholarship was awarded in an institution of higher education, and is not engaging in gainful employment other than employment approved by the Administrator by or pursuant to regulation.

"(6) The Administrator shall by regulation provide that any person awarded a scholarship under this section shall agree in writing to enter and remain in an occupation involving the design, operation, or maintenance of treatment works for such period after completion of his course of studies as the Administrator determines appropriate.

"DEFINITIONS AND AUTHORIZATIONS

"SEC. 112. (a) As used in sections 109 through 112 of this Act—
"(1) The term 'institution of higher education' means an educational institution described in the first sentence of section 1201 of the Higher Education Act of 1965 (other than an institution of any agency of the United States) which is accredited by a nationally recognized accrediting agency or association approved by the Administrator for this purpose. For purposes of this subsection, the Administrator shall publish a list of nationally recognized accrediting agencies or associations which he determines to be reliable authority as to the quality of training offered. 79 Stat. 1269; 82 Stat. 1042, 1050. 20 USC 1141.

"(2) The term 'academic year' means an academic year or its equivalent, as determined by the Administrator.

"(b) The Administrator shall annually report his activities under sections 109 through 112 of this Act, including recommendations for needed revisions in the provisions thereof. Annual report.

"(c) There are authorized to be appropriated $25,000,000 per fiscal year for the fiscal years ending June 30, 1973, and June 30, 1974, to carry out sections 109 through 112 of this Act.

"ALASKA VILLAGE DEMONSTRATION PROJECTS

"SEC. 113. (a) The Administrator is authorized to enter into agreements with the State of Alaska to carry out one or more projects to demonstrate methods to provide for central community facilities for safe water and elimination or control of pollution in those native villages of Alaska without such facilities. Such project shall include provisions for community safe water supply systems, toilets, bathing and laundry facilities, sewage disposal facilities, and other similar facilities, and educational and informational facilities and programs relating to health and hygiene. Such demonstration projects shall be for the further purpose of developing preliminary plans for providing such safe water and such elimination or control of pollution for all native villages in such State.

"(b) In carrying out this section the Administrator shall cooperate with the Secretary of Health, Education, and Welfare for the purpose of utilizing such of the personnel and facilities of that Department as may be appropriate. HEW, cooperation.

139

"(c) The Administrator shall report to Congress not later than July 1, 1973, the results of the demonstration projects authorized by this section together with his recommendations, including any necessary legislation, relating to the establishment of a statewide program.

"(d) There is authorized to be appropriated not to exceed $2,000,000 to carry out this section.

"LAKE TAHOE STUDY

"SEC. 114. (a) The Administrator, in consultation with the Tahoe Regional Planning Agency, the Secretary of Agriculture, other Federal agencies. representatives of State and local governments, and members of the public, shall conduct a thorough and complete study on the adequacy of and need for extending Federal oversight and control in order to preserve the fragile ecology of Lake Tahoe.

"(b) Such study shall include an examination of the interrelationships and responsibilities of the various agencies of the Federal Government and State and local governments with a view to establishing the necessity for redefinition of legal and other arrangements between these various governments, and making specific legislative recommendations to Congress. Such study shall consider the effect of various actions in terms of their environmental impact on the Tahoe Basin, treated as an ecosystem.

"(c) The Administrator shall report on such study to Congress not later than one year after the date of enactment of this subsection.

"(d) There is authorized to be appropriated to carry out this section not to exceed $500,000.

"IN-PLACE TOXIC POLLUTANTS

"SEC. 115. The Administrator is directed to identify the location of in-place pollutants with emphasis on toxic pollutants in harbors and navigable waterways and is authorized, acting through the Secretary of the Army, to make contracts for the removal and appropriate disposal of such materials from critical port and harbor areas. There is authorized to be appropriated $15,000,000 to carry out the provisions of this section, which sum shall be available until expended.

"TITLE II—GRANTS FOR CONSTRUCTION OF TREATMENT WORKS

"PURPOSE

"SEC. 201. (a) It is the purpose of this title to require and to assist the development and implementation of waste treatment management plans and practices which will achieve the goals of this Act.

"(b) Waste treatment management plans and practices shall provide for the application of the best practicable waste treatment technology before any discharge into receiving waters, including reclaiming and recycling of water, and confined disposal of pollutants so they will not migrate to cause water or other environmental pollution and shall provide for consideration of advanced waste treatment techniques.

"(c) To the extent practicable, waste treatment management shall be on an areawide basis and provide control or treatment of all point and nonpoint sources of pollution, including in place or accumulated pollution sources.

"(d) The Administrator shall encourage waste treatment management which results in the construction of revenue producing facilities providing for—

"(1) the recycling of potential sewage pollutants through the production of agriculture, silviculture, or aquaculture products, or any combination thereof;

"(2) the confined and contained disposal of pollutants not recycled;

"(3) the reclamation of wastewater; and

"(4) the ultimate disposal of sludge in a manner that will not result in environmental hazards.

"(e) The Administrator shall encourage waste treatment management which results in integrating facilities for sewage treatment and recycling with facilities to treat, dispose of, or utilize other industrial and municipal wastes, including but not limited to solid waste and waste heat and thermal discharges. Such integrated facilities shall be designed and operated to produce revenues in excess of capital and operation and maintenance costs and such revenues shall be used by the designated regional management agency to aid in financing other environmental improvement programs.

"(f) The Administrator shall encourage waste treatment management which combines 'open space' and recreational considerations with such management.

"(g)(1) The Administrator is authorized to make grants to any State, municipality, or intermunicipal or interstate agency for the construction of publicly owned treatment works.

"(2) The Administrator shall not make grants from funds authorized for any fiscal year beginning after June 30, 1974, to any State, municipality, or intermunicipal or interstate agency for the erection, building, acquisition, alteration, remodeling, improvement. or extension of treatment works unless the grant applicant has satisfactorily demonstrated to the Administrator that— Conditions.

"(A) alternative waste management techniques have been studied and evaluated and the works proposed for grant assistance will provide for the application of the best practicable waste treatment technology over the life of the works consistent with the purposes of this title; and

"(B) as appropriate, the works proposed for grant assistance will take into account and allow to the extent practicable the application of technology at a later date which will provide for the reclaiming or recycling of water or otherwise eliminate the discharge of pollutants.

"(3) The Administrator shall not approve any grant after July 1, 1973, for treatment works under this section unless the applicant shows to the satisfaction of the Administrator that each sewer collection system discharging into such treatment works is not subject to excessive infiltration.

"(4) The Administrator is authorized to make grants to applicants for treatment works grants under this section for such sewer system evaluation studies as may be necessary to carry out the requirements of paragraph (3) of this subsection. Such grants shall be made in accordance with rules and regulations promulgated by the Administrator. Initial rules and regulations shall be promulgated under this paragraph not later than 120 days after the date of enactment of the Federal Water Pollution Control Act Amendments of 1972. Rules and regulations.

"FEDERAL SHARE

"SEC. 202. (a) The amount of any grant for treatment works made under this Act from funds authorized for any fiscal year beginning after June 30, 1971, shall be 75 per centum of the cost of construction

141

thereof (as approved by the Administrator). Any grant (other than for reimbursement) made prior to the date of enactment of the Federal Water Pollution Control Act Amendments of 1972 from any funds authorized for any fiscal year beginning after June 30, 1971, shall, upon the request of the applicant, be increased to the applicable percentage under this section.

"(b) The amount of the grant for any project approved by the Administrator after January 1, 1971, and before July 1, 1971, for the construction of treatment works, the actual erection, building or acquisition of which was not commenced prior to July 1, 1971, shall, upon the request of the applicant, be increased to the applicable percentage under subsection (a) of this section for grants for treatment works from funds for fiscal years beginning after June 30, 1971, with respect to the cost of such actual erection, building, or acquisition. Such increased amount shall be paid from any funds allocated to the State in which the treatment works is located without regard to the fiscal year for which such funds were authorized. Such increased amount shall be paid for such project only if—

"(1) a sewage collection system that is a part of the same total waste treatment system as the treatment works for which such grant was approved is under construction or is to be constructed for use in conjunction with such treatment works, and if the cost of such sewage collection system exceeds the cost of such treatment works, and

"(2) the State water pollution control agency or other appropriate State authority certifies that the quantity of available ground water will be insufficient, inadequate, or unsuitable for public use, including the ecological preservation and recreational use of surface water bodies, unless effluents from publicly-owned treatment works after adequate treatment are returned to the ground water consistent with acceptable technological standards.

"PLANS, SPECIFICATIONS, ESTIMATES, AND PAYMENTS

"Sec. 203. (a) Each applicant for a grant shall submit to the Administrator for his approval, plans, specifications, and estimates for each proposed project for the construction of treatment works for which a grant is applied for under section 201(g)(1) from funds allotted to the State under section 205 and which otherwise meets the requirements of this Act. The Administrator shall act upon such plans, specifications, and estimates as soon as practicable after the same have been submitted, and his approval of any such plans, specifications, and estimates shall be deemed a contractual obligation of the United States for the payment of its proportional contribution to such project.

Limitation.

"(b) The Administrator shall, from time to time as the work progresses, make payments to the recipient of a grant for costs of construction incurred on a project. These payments shall at no time exceed the Federal share of the cost of construction incurred to the date of the voucher covering such payment plus the Federal share of the value of the materials which have been stockpiled in the vicinity of such construction in conformity to plans and specifications for the project.

"(c) After completion of a project and approval of the final voucher by the Administrator, he shall pay out of the appropriate sums the unpaid balance of the Federal share payable on account of such project.

"LIMITATIONS AND CONDITIONS

"Sec. 204. (a) Before approving grants for any project for any treatment works under section 201(g)(1) the Administrator shall determine—

"(1) that such works are included in any applicable areawide waste treatment management plan developed under section 208 of this Act;

"(2) that such works are in comformity with any applicable State plan under section 303(e) of this Act;

"(3) that such works have been certified by the appropriate State water pollution control agency as entitled to priority over such other works in the State in accordance with any applicable State plan under section 303(e) of this Act;

"(4) that the applicant proposing to construct such works agrees to pay the non-Federal costs of such works and has made adequate provisions satisfactory to the Administrator for assuring proper and efficient operation, including the employment of trained management and operations personnel, and the maintenance of such works in accordance with a plan of operation approved by the State water pollution control agency or, as appropriate, the interstate agency, after construction thereof;

"(5) that the size and capacity of such works relate directly to the needs to be served by such works, including sufficient reserve capacity. The amount of reserve capacity provided shall be approved by the Administrator on the basis of a comparison of the cost of constructing such reserves as a part of the works to be funded and the anticipated cost of providing expanded capacity at a date when such capacity will be required;

"(6) that no specification for bids in connection with such works shall be written in such a manner as to contain proprietary, exclusionary, or discriminatory requirements other than those based upon performance, unless such requirements are necessary to test or demonstrate a specific thing or to provide for necessary interchangeability of parts and equipment, or at least two brand names or trade names of comparable quality or utility are listed and are followed by the words 'or equal'.

"(b)(1) Notwithstanding any other provision of this title, the Administrator shall not approve any grant for any treatment works under section 201(g)(1) after March 1, 1973, unless he shall first have determined that the applicant (A) has adopted or will adopt a system of charges to assure that each recipient of waste treatment services within the applicant's jurisdiction, as determined by the Administrator, will pay its proportionate share of the costs of operation and maintenance (including replacement) of any waste treatment services provided by the applicant; (B) has made provision for the payment to such applicant by the industrial users of the treatment works, of that portion of the cost of construction of such treatment works (as determined by the Administrator) which is allocable to the treatment of such industrial wastes to the extent attributable to the Federal share of the cost of construction; and (C) has legal, institutional, managerial, and financial capability to insure adequate construction, operation, and maintenance of treatment works throughout the applicant's jurisdiction, as determined by the Administrator.

"(2) The Administrator shall, within one hundred and eighty days after the date of enactment of the Federal Water Pollution Control Act Amendments of 1972, and after consultation with appropriate State, interstate, municipal, and intermunicipal agencies, issue guidelines applicable to payment of waste treatment costs by industrial and nonindustrial recipients of waste treatment services which shall establish (A) classes of users of such services, including categories of industrial users; (B) criteria against which to determine the adequacy of charges imposed on classes and categories of users reflecting all factors that influence the cost of waste treatment, including strength, volume, and delivery flow rate characteristics of waste; and (C) model systems and rates of user charges typical of various treatment works serving municipal-industrial communities.

"(3) The grantee shall retain an amount of the revenues derived from the payment of costs by industrial users of waste treatment services, to the extent costs are attributable to the Federal share of eligible project costs provided pursuant to this title as determined by the Administrator, equal to (A) the amount of the non-Federal cost of such project paid by the grantee plus (B) the amount, determined in accordance with regulations promulgated by the Administrator, necessary for future expansion and reconstruction of the project, except that such retained amount shall not exceed 50 per centum of such revenues from such project. All revenues from such project not retained by the grantee shall be deposited by the Administrator in the Treasury as miscellaneous receipts. That portion of the revenues retained by the grantee attributable to clause (B) of the first sentence of this paragraph, together with any interest thereon shall be used solely for the purposes of future expansion and reconstruction of the project.

"(4) Approval by the Administrator of a grant to an interstate agency established by interstate compact for any treatment works shall satisfy any other requirement that such works be authorized by Act of Congress.

"ALLOTMENT

"SEC. 205. (a) Sums authorized to be appropriated pursuant to section 207 for each fiscal year beginning after June 30, 1972, shall be allotted by the Administrator not later than the January 1st immediately preceding the beginning of the fiscal year for which authorized, except that the allotment for fiscal year 1973 shall be made not later than 30 days after the date of enactment of the Federal Water Pollution Control Act Amendments of 1972. Such sums shall be allotted among the States by the Administrator in accordance with regulations promulgated by him, in the ratio that the estimated cost of constructing all needed publicly owned treatment works in each State bears to the estimated cost of construction of all needed publicly owned treatment works in all of the States. For the fiscal years ending June 30, 1973, and June 30, 1974, such ratio shall be determined on the basis of table III of House Public Works Committee Print No. 92–50. Allotments for fiscal years which begin after the fiscal year ending June 30, 1974, shall be made only in accordance with a revised cost estimate made and submitted to Congress in accordance with section 516(b) of this Act and only after such revised cost estimate shall have been approved by law specifically enacted hereafter.

"(b)(1) Any sums allotted to a State under subsection (a) shall be available for obligation under section 203 on and after the date of such allotment. Such sums shall continue available for obligation in such State for a period of one year after the close of the fiscal year for which such sums are authorized. Any amounts so allotted which are not obligated by the end of such one-year period shall be immediately reallotted by the Administrator, in accordance with regulations promulgated by him, generally on the basis of the ratio used in making the last allotment of sums under this section. Such reallotted sums shall be added to the last allotments made to the States. Any sum made available to a State by reallotment under this subsection shall be in addition to any funds otherwise allotted to such State for grants under this title during any fiscal year.

144

"(2) Any sums which have been obligated under section 203 and which are released by the payment of the final voucher for the project shall be immediately credited to the State to which such sums were last allotted. Such released sums shall be added to the amounts last allotted to such State and shall be immediately available for obligation in the same manner and to the same extent as such last allotment.

"REIMBURSEMENT AND ADVANCED CONSTRUCTION

"SEC. 206. (a) Any publicly owned treatment works in a State on which construction was initiated after June 30, 1966, but before July 1, 1972, which was approved by the appropriate State water pollution control agency and which the Administrator finds meets the requirements of section 8 of this Act in effect at the time of the initiation of construction shall be reimbursed a total amount equal to the difference between the amount of Federal financial assistance, if any, received under such section 8 for such project and 50 per centum of the cost of such project, or 55 per centum of the project cost where the Administrator also determines that such treatment works was constructed in conformity with a comprehensive metropolitan treatment plan as described in section 8(f) of the Federal Water Pollution Control Act as in effect immediately prior to the date of enactment of the Federal Water Pollution Control Act Amendments of 1972. Nothing in this subsection shall result in any such works receiving Federal grants from all sources in excess of 80 per centum of the cost of such project.

79 Stat. 907.
33 USC 1158.

"(b) Any publicly owned treatment works constructed with or eligible for Federal financial assistance under this Act in a State between June 30, 1956, and June 30, 1966, which was approved by the State water pollution control agency and which the Administrator finds meets the requirements of section 8 of this Act prior to the date of enactment of the Federal Water Pollution Control Act Amendments of 1972 but which was constructed without assistance under such section 8 or which received such assistance in an amount less than 30 per centum of the cost of such project shall qualify for payments and reimbursement of State or local funds used for such project from sums allocated to such State under this section in an amount which shall not exceed the difference between the amount of such assistance, if any, received for such project and 30 per centum of the cost of such project.

"(c) No publicly owned treatment works shall receive any payment or reimbursement under subsection (a) or (b) of this section unless an application for such assistance is filed with the Administrator within the one year period which begins on the date of enactment of the Federal Water Pollution Control Act Amendments of 1972. Any application filed within such one year period may be revised from time to time, as may be necessary.

"(d) The Administrator shall allocate to each qualified project under subsection (a) of this section each fiscal year for which funds are appropriated under subsection (e) of this section an amount which bears the same ratio to the unpaid balance of the reimbursement due such project as the total of such funds for such year bears to the total unpaid balance of reimbursement due all such approved projects on the date of enactment of such appropriation. The Administrator shall allocate to each qualified project under subsection (b) of this section each fiscal year for which funds are appropriated under subsection (e) of this section an amount which bears the same ratio to the unpaid balance of the reimbursement due such project as the total of such funds

145

for such year bears to the total unpaid balance of reimbursement due all such approved projects on the date of enactment of such appropriation.

"(e) There is authorized to be appropriated to carry out subsection {a} of this section not to exceed $2,000,000,000 and, to carry out subsection (b) of this section, not to exceed $750,000,000. The authorizations contained in this subsection shall be the sole source of funds for reimbursements authorized by this section.

"(f)(1) In any case where all funds allotted to a State under this title have been obligated under section 203 of this Act, and there is construction of any treatment works project without the aid of Federal funds and in accordance with all procedures and all requirements applicable to treatment works projects, except those procedures and requirements which limit construction of projects to those constructed with the aid of previously allotted Federal funds, the Administrator, upon his approval of an application made under this subsection therefor, is authorized to pay the Federal share of the cost of construction of such project when additional funds are allotted to the State under this title if prior to the construction of the project the Administrator approves plans, specifications, and estimates therefor in the same manner as other treatment works projects. The Administrator may not approve an application under this subsection unless an authorization is in effect for the future fiscal year for which the application requests payment, which authorization will insure such payment without exceeding the State's expected allotment from such authorization.

"(2) In determining the allotment for any fiscal year under this title, any treatment works project constructed in accordance with this section and without the aid of Federal funds shall not be considered completed until an application under the provisions of this subsection with respect to such project has been approved by the Administrator, or the availability of funds from which this project is eligible for reimbursement has expired, whichever first occurs.

"AUTHORIZATION

"SEC. 207. There is authorized to be appropriated to carry out this title, other than sections 208 and 209, for the fiscal year ending June 30, 1973, not to exceed $5,000,000,000, for the fiscal year ending June 30, 1974, not to exceed $6,000,000,000, and for the fiscal year ending June 30, 1975, not to exceed $7,000,000,000.

"AREAWIDE WASTE TREATMENT MANAGEMENT

"SEC. 208. (a) For the purpose of encouraging and facilitating the development and implementation of areawide waste treatment management plans—

Guidelines, publication.

"(1) The Administrator, within ninety days after the date of enactment of this Act and after consultation with appropriate Federal, State, and local authorities, shall by regulation publish guidelines for the identification of those areas which, as a result of urban-industrial concentrations or other factors, have substantial water quality control problems.

Boundaries; planning agencies.

"(2) The Governor of each State, within sixty days after publication of the guidelines issued pursuant to paragraph (1) of this subsection, shall identify each area within the State which, as a result of urban-industrial concentrations or other factors, has substantial water quality control problems. Not later than one hundred and twenty days following such identification and after consultation with appropriate elected and other officials of local

146

governments having jurisdiction in such areas, the Governor shall designate (A) the boundaries of each such area, and (B) a single representative organization, including elected officials from local governments or their designees, capable of developing effective areawide waste treatment management plans for such area. The Governor may in the same manner at any later time identify any additional area (or modify an existing area) for which he determines areawide waste treatment management to be appropriate, designate the boundaries of such area, and designate an organization capable of developing effective areawide waste treatment management plans for such area.

"(3) With respect to any area which, pursuant to the guidelines published under paragraph (1) of this subsection, is located in two or more States, the Governors of the respective States shall consult and cooperate in carrying out the provisions of paragraph (2), with a view toward designating the boundaries of the interstate area having common water quality control problems and for which areawide waste treatment management plans would be most effective, and toward designating, within one hundred and eighty days after publication of guidelines issued pursuant to paragraph (1) of this subsection, of a single representative organization capable of developing effective areawide waste treatment management plans for such area.

"(4) If a Governor does not act, either by designating or determining not to make a designation under paragraph (2) of this subsection, within the time required by such paragraph, or if, in the case of an interstate area, the Governors of the States involved do not designate a planning organization within the time required by paragraph (3) of this subsection, the chief elected officials of local governments within an area may by agreement designate (A) the boundaries for such an area, and (B) a single representative organization including elected officials from such local governments, or their designees, capable of developing an areawide waste treatment management plan for such area.

"(5) Existing regional agencies may be designated under paragraphs (2), (3), and (4) of this subsection.

"(6) The State shall act as a planning agency for all portions of such State which are not designated under paragraphs (2), (3), or (4) of this subsection.

"(7) Designations under this subsection shall be subject to the approval of the Administrator.

"(b)(1) Not later than one year after the date of designation of any organization under subsection (a) of this section such organization shall have in operation a continuing areawide waste treatment management planning process consistent with section 201 of this Act. Plans prepared in accordance with this process shall contain alternatives for waste treatment management, and be applicable to all wastes generated within the area involved. The initial plan prepared in accordance with such process shall be certified by the Governor and submitted to the Administrator not later than two years after the planning process is in operation.

Waste treatment management planning process.

"(2) Any plan prepared under such process shall include, but not be limited to—

"(A) the identification of treatment works necessary to meet the anticipated municipal and industrial waste treatment needs of the area over a twenty-year period, annually updated (including an analysis of alternative waste treatment systems), including any requirements for the acquisition of land for treatment pur-

147

poses; the necessary waste water collection and urban storm water runoff systems; and a program to provide the necessary financial arrangements for the development of such treatment works;

"(B) the establishment of construction priorities for such treatment works and time schedules for the initiation and completion of all treatment works;

"(C) the establishment of a regulatory program to—

"(i) implement the waste treatment management requirements of section 201(c),

"(ii) regulate the location, modification, and construction of any facilities within such area which may result in any discharge in such area, and

"(iii) assure that any industrial or commercial wastes discharged into any treatment works in such area meet applicable pretreatment requirements;

"(D) the identification of those agencies necessary to construct, operate, and maintain all facilities required by the plan and otherwise to carry out the plan;

"(E) the identification of the measures necessary to carry out the plan (including financing), the period of time necessary to carry out the plan, the costs of carrying out the plan within such time, and the economic, social, and environmental impact of carrying out the plan within such time;

"(F) a process to (i) identify, if appropriate, agriculturally and silviculturally related nonpoint sources of pollution, including runoff from manure disposal areas, and from land used for livestock and crop production, and (ii) set forth procedures and methods (including land use requirements) to control to the extent feasible such sources;

"(G) a process to (i) identify, if appropriate, mine-related sources of pollution including new, current, and abandoned surface and underground mine runoff, and (ii) set forth procedures and methods (including land use requirements) to control to the extent feasible such sources;

"(H) a process to (i) identify construction activity related sources of pollution, and (ii) set forth procedures and methods (including land use requirements) to control to the extent feasible such sources;

"(I) a process to (i) identify, if appropriate, salt water intrusion into rivers, lakes, and estuaries resulting from reduction of fresh water flow from any cause, including irrigation, obstruction, ground water extraction, and diversion, and (ii) set forth procedures and methods to control such intrusion to the extent feasible where such procedures and methods are otherwise a part of the waste treatment management plan;

"(J) a process to control the disposition of all residual waste generated in such area which could affect water quality; and

"(K) a process to control the disposal of pollutants on land or in subsurface excavations within such area to protect ground and surface water quality.

Annual certification. "(3) Areawide waste treatment management plans shall be certified annually by the Governor or his designee (or Governors or their designees, where more than one State is involved) as being consistent with applicable basin plans and such areawide waste treatment management plans shall be submitted to the Administrator for his approval.

"(4) Whenever the Governor of any State determines (and notifies the Administrator) that consistency with a statewide regulatory program under section 303 so requires, the requirements of clauses (F)

through (K) of paragraph (2) of this subsection shall be developed and submitted by the Governor to the Administrator for application to all regions within such State.

"(c)(1) The Governor of each State, in consultation with the planning agency designated under subsection (a) of this section, at the time a plan is submitted to the Administrator, shall designate one or more waste treatment management agencies (which may be an existing or newly created local, regional, or State agency or political subdivision) for each area designated under subsection (a) of this section and submit such designations to the Administrator. Regional operating agencies, designation.

"(2) The Administrator shall accept any such designation, unless, within 120 days of such designation, he finds that the designated management agency (or agencies) does not have adequate authority—

"(A) to carry out appropriate portions of an areawide waste treatment management plan developed under subsection (b) of this section;

"(B) to manage effectively waste treatment works and related facilities serving such area in conformance with any plan required by subsection (b) of this section;

"(C) directly or by contract, to design and construct new works, and to operate and maintain new and existing works as required by any plan developed pursuant to subsection (b) of this section;

"(D) to accept and utilize grants, or other funds from any source, for waste treatment management purposes;

"(E) to raise revenues, including the assessment of waste treatment charges;

"(F) to incur short- and long-term indebtedness;

"(G) to assure in implementation of an areawide waste treatment management plan that each participating community pays its proportionate share of treatment costs;

"(H) to refuse to receive any wastes from any municipality or subdivision thereof, which does not comply with any provisions of an approved plan under this section applicable to such area; and

"(I) to accept for treatment industrial wastes.

"(d) After a waste treatment management agency having the authority required by subsection (c) has been designated under such subsection for an area and a plan for such area has been approved under subsection (b) of this section, the Administrator shall not make any grant for construction of a publicly owned treatment works under section 201(g)(1) within such area except to such designated agency and for works in conformity with such plan.

"(e) No permit under section 402 of this Act shall be issued for any point source which is in conflict with a plan approved pursuant to subsection (b) of this section.

"(f)(1) The Administrator shall make grants to any agency designated under subsection (a) of this section for payment of the reasonable costs of developing and operating a continuing areawide waste treatment management planning process under subsection (b) of this section. Grants.

"(2) The amount granted to any agency under paragraph (1) of this subsection shall be 100 per centum of the costs of developing and operating a continuing areawide waste treatment management planning process under subsection (b) of this section for each of the fiscal years ending on June 30, 1973, June 30, 1974, and June 30, 1975, and shall not exceed 75 per centum of such costs in each succeeding fiscal year.

"(3) Each applicant for a grant under this subsection shall submit to the Administrator for his approval each proposal for which a grant is applied for under this subsection. The Administrator shall act upon such proposal as soon as practicable after it has been submitted. and his approval of that proposal shall be deemed a contractual obligation of the United States for the payment of its contribution to such proposal. There is authorizd to be appropriated to carry out this subsection not to exceed $50,000,000 for the fiscal year ending June 30, 1973, not to exceed $100,000,000 for the fiscal year ending June 30, 1974, and not to exceed $150,000,000 for the fiscal year ending June 30, 1975.

Appropriation.

Technical
assistance.

"(g) The Administrator is authorized, upon request of the Governor or the designated planning agency, and without reimbursement, to consult with, and provide technical assistance to, any agency designated under subsection (a) of this section in the development of areawide waste treatment management plans under subsection (b) of this section.

"(h)(1) The Secretary of the Army, acting through the Chief of Engineers, in cooperation with the Administrator is authorized and directed, upon request of the Governor or the designated planning organization, to consult with, and provide technical assistance to, any agency designed under subsection (a) of this section in developing and operating a continuing areawide waste treatment management planning process under subsection (b) of this section.

Appropriation.

"(2) There is authorized to be appropriated to the Secretary of the Army, to carry out this subsection, not to exceed $50,000,000 per fiscal year for the fiscal years ending June 30, 1973, and June 30, 1974.

"BASIN PLANNING

"SEC. 209 (a) The President, acting through the Water Resources Council, shall, as soon as practicable, prepare a Level B plan under the Water Resources Planning Act for all basins in the United States. All such plans shall be completed not later than January 1, 1980, except that priority in the preparation of such plans shall be given to those basins and portions thereof which are within those areas designated under paragraphs (2), (3), and (4) of subsection (a) of section 208 of this Act.

79 Stat. 244.
42 USC 1962
note.

Annual report
to Congress.

"(b) The President, acting through the Water Resources Council, shall report annually to Congress on progress being made in carrying out this section. The first such report shall be submitted not later than January 31, 1973.

Appropriation.

"(c) There is authorized to be appropriated to carry out this section not to exceed $200,000,000.

"ANNUAL SURVEY

"SEC. 210. The Administrator shall annually make a survey to determine the efficiency of the operation and maintenance of treatment works constructed with grants made under this Act. as compared to the efficiency planned at the time the grant was made. The results of such annual survey shall be included in the report required under section 516(a) of this Act.

"SEWAGE COLLECTION SYSTEMS

"SEC. 211. No grant shall be made for a sewage collection system under this title unless such grant (1) is for replacement or major rehabilitation of an existing collection system and is necessary to the total integrity and performance of the waste treatment works servicing such community, or (2) is for a new collection system in an existing community with sufficient existing or planned capacity adequately to treat such collected sewage and is consistent with section 201 of this Act.

"DEFINITIONS

"SEC. 212. As used in this title—

"(1) The term 'construction' means any one or more of the following: preliminary planning to determine the feasibility of treatment works, engineering, architectural, legal, fiscal, or economic investigations or studies, surveys, designs, plans, working drawings, specifications, procedures, or other necessary actions, erection, building, acquisition, alteration, remodeling, improvement, or extension of treatment works, or the inspection or supervision of any of the foregoing items.

"(2) (A) The term 'treatment works' means any devices and systems used in the storage, treatment, recycling, and reclamation of municipal sewage or industrial wastes of a liquid nature to implement section 201 of this Act, or necessary to recycle or reuse water at the most economical cost over the estimated life of the works, including intercepting sewers, outfall sewers, sewage collection systems, pumping, power, and other equipment, and their appurtenances; extensions, improvements, remodeling, additions, and alterations thereof; elements essential to provide a reliable recycled supply such as standby treatment units and clear well facilities; and any works, including site acquisition of the land that will be an integral part of the treatment process or is used for ultimate disposal of residues resulting from such treatment.

"(B) In addition to the definition contained in subparagraph (A) of this paragraph, 'treatment works' means any other method or system for preventing, abating, reducing, storing, treating, separating, or disposing of municipal waste, including storm water runoff, or industrial waste, including waste in combined storm water and sanitary sewer systems. Any application for construction grants which includes wholly or in part such methods or systems shall, in accordance with guidelines published by the Administrator pursuant to subparagraph (C) of this paragraph, contain adequate data and analysis demonstrating such proposal to be, over the life of such works, the most cost efficient alternative to comply with sections 301 or 302 of this Act, or the requirements of section 201 of this Act.

"(C) For the purposes of subparagraph (B) of this paragraph, the Administrator shall, within one hundred and eighty days after the date of enactment of this title, publish and thereafter revise no less often than annually, guidelines for the evaluation of methods, including cost-effective analysis, described in subparagraph (B) of this paragraph. *Methods, evaluation guidelines, publication.*

"(3) The term 'replacement' as used in this title means those expenditures for obtaining and installing equipment, accessories, or appurtenances during the useful life of the treatment works necessary to maintain the capacity and performance for which such works are designed and constructed.

"TITLE III—STANDARDS AND ENFORCEMENT

"EFFLUENT LIMITATIONS

"SEC. 301. (a) Except as in compliance with this section and sections 302, 306, 307, 318, 402, and 404 of this Act, the discharge of any pollutant by any person shall be unlawful.

"(b) In order to carry out the objective of this Act there shall be achieved—

151

"(1) (A) not later than July 1, 1977, effluent limitations for point sources, other than publicly owned treatment works, (i) which shall require the application of the best practicable control technology currently available as defined by the Administrator pursuant to section 304(b) of this Act, or (ii) in the case of a discharge into a publicly owned treatment works which meets the requirements of subparagraph (B) of this paragraph, which shall require compliance with any applicable pretreatment requirements and any requirements under section 307 of this Act; and

"(B) for publicly owned treatment works in existence on July 1, 1977, or approved pursuant to section 203 of this Act prior to June 30, 1974 (for which construction must be completed within four years of approval), effluent limitations based upon secondary treatment as defined by the Administrator pursuant to section 304(d) (1) of this Act; or,

"(C) not later than July 1, 1977, any more stringent limitation, including those necessary to meet water quality standards, treatment standards, or schedules of compliance, established pursuant to any State law or regulations (under authority preserved by section 510) or any other Federal law or regulation, or required to implement any applicable water quality standard established pursuant to this Act.

"(2) (A) not later than July 1, 1983, effluent limitations for categories and classes of point sources, other than publicly owned treatment works, which (i) shall require application of the best available technology economically achievable for such category or class, which will result in reasonable further progress toward the national goal of eliminating the discharge of all pollutants, as determined in accordance with regulations issued by the Administrator pursuant to section 304(b) (2) of this Act, which such effluent limitations shall require the elimination of discharges of all pollutants if the Administrator finds, on the basis of information available to him (including information developed pursuant to section 315), that such elimination is technologically and economically achievable for a category or class of point sources as determined in accordance with regulations issued by the Administrator pursuant to section 304(b) (2) of this Act, or (ii) in the case of the introduction of a pollutant into a publicly owned treatment works which meets the requirements of subparagraph (B) of this paragraph, shall require compliance with any applicable pretreatment requirements and any other requirement under section 307 of this Act; and

"(B) not later than July 1, 1983, compliance by all publicly owned treatment works with the requirements set forth in section 201(g) (2) (A) of this Act.

"(c) The Administrator may modify the requirements of subsection (b) (2) (A) of this section with respect to any point source for which a permit application is filed after July 1, 1977, upon a showing by the owner or operator of such point source satisfactory to the Administrator that such modified requirements (1) will represent the maximum use of technology within the economic capability of the owner or operator; and (2) will result in reasonable further progress toward the elimination of the discharge of pollutants.

"(d) Any effluent limitation required by paragraph (2) of subsection (b) of this section shall be reviewed at least every five years and, if appropriate, revised pursuant to the procedure established under such paragraph.

"(e) Effluent limitations established pursuant to this section or section 302 of this Act shall be applied to all point sources of discharge of pollutants in accordance with the provisions of this Act.

152

"(f) Notwithstanding any other provisions of this Act it shall be unlawful to discharge any radiological, chemical, or biological warfare agent or high-level radioactive waste into the navigable waters.

"WATER QUALITY RELATED EFFLUENT LIMITATIONS

"SEC. 302. (a) Whenever, in the judgment of the Administrator, discharges of pollutants from a point source or group of point sources, with the application of effluent limitations required under section 301 (b)(2) of this Act, would interfere with the attainment or maintenance of that water quality in a specific portion of the navigable waters which shall assure protection of public water supplies, agricultural and industrial uses, and the protection and propagation of a balanced population of shellfish, fish and wildlife, and allow recreational activities in and on the water, effluent limitations (including alternative effluent control strategies) for such point source or sources shall be established which can reasonably be expected to contribute to the attainment or maintenance of such water quality.

"(b)(1) Prior to establishment of any effluent limitation pursuant to subsection (a) of this section, the Administrator shall issue notice of intent to establish such limitation and within ninety days of such notice hold a public hearing to determine the relationship of the economic and social costs of achieving any such limitation or limitations, including any economic or social dislocation in the affected community or communities, to the social and economic benefits to be obtained (including the attainment of the objective of this Act) and to determine whether or not such effluent limitations can be implemented with available technology or other alternative control strategies. *Public hearing.*

"(2) If a person affected by such limitation demonstrates at such hearing that (whether or not such technology or other alternative control strategies are available) there is no reasonable relationship between the economic and social costs and the benefits to be obtained (including attainment of the objective of this Act), such limitation shall not become effective and the Administrator shall adjust such limitation as it applies to such person.

"(c) The establishment of effluent limitations under this section shall not operate to delay the application of any effluent limitation established under section 301 of this Act.

"WATER QUALITY STANDARDS AND IMPLEMENTATION PLANS

"SEC. 303. (a)(1) In order to carry out the purpose of this Act, any water quality standard applicable to interstate waters which was adopted by any State and submitted to, and approved by, or is awaiting approval by, the Administrator pursuant to this Act as in effect immediately prior to the date of enactment of the Federal Water Pollution Control Act Amendments of 1972, shall remain in effect unless the Administrator determined that such standard is not consistent with the applicable requirements of this Act as in effect immediately prior to the date of enactment of the Federal Water Pollution Control Act Amendments of 1972. If the Administrator makes such a determination he shall, within three months after the date of enactment of the Federal Water Pollution Control Act Amendments of 1972, notify the State and specify the changes needed to meet such requirements. If such changes are not adopted by the State within ninety days after the date of such notification, the Administrator shall promulgate such changes in accordance with subsection (b) of this section.

"(2) Any State which, before the date of enactment of the Federal Water Pollution Control Act Amendments of 1972, has adopted, pursuant to its own law, water quality standards applicable to intrastate waters shall submit such standards to the Administrator within thirty days after the date of enactment of the Federal Water Pollution Control Act Amendments of 1972. Each such standard shall remain in effect, in the same manner and to the same extent as any other water quality standard established under this Act unless the Administrator determines that such standard is inconsistent with the applicable requirements of this Act as in effect immediately prior to the date of enactment of the Federal Water Pollution Control Act Amendments of 1972. If the Administrator makes such a determination he shall not later than the one hundred and twentieth day after the date of submission of such standards, notify the State and specify the changes needed to meet such requirements. If such changes are not adopted by the State within ninety days after such notification, the Administrator shall promulgate such changes in accordance with subsection (b) of this section.

"(3)(A) Any State which prior to the date of enactment of the Federal Water Pollution Control Act Amendments of 1972 has not adopted pursuant to its own laws water quality standards applicable to intrastate waters shall, not later than one hundred and eighty days after the date of enactment of the Federal Water Pollution Control Act Amendments of 1972, adopt and submit such standards to the Administrator.

"(B) If the Administrator determines that any such standards are consistent with the applicable requirements of this Act as in effect immediately prior to the date of enactment of the Federal Water Pollution Control Act Amendments of 1972, he shall approve such standards.

"(C) If the Administrator determines that any such standards are not consistent with the applicable requirements of this Act as in effect immediately prior to the date of enactment of the Federal Water Pollution Control Act Amendments of 1972, he shall, not later than the ninetieth day after the date of submission of such standards, notify the State and specify the changes to meet such requirements. If such changes are not adopted by the State within ninety days after the date of notification, the Administrator shall promulgate such standards pursuant to subsection (b) of this section.

Proposed regulations, publication. "(b)(1) The Administrator shall promptly prepare and publish proposed regulations setting forth water quality standards for a State in accordance with the applicable requirements of this Act as in effect immediately prior to the date of enactment of the Federal Water Pollution Control Act Amendments of 1972, if—

"(A) the State fails to submit water quality standards within the times prescribed in subsection (a) of this section,

"(B) a water quality standard submitted by such State under subsection (a) of this section is determined by the Administrator not to be consistent with the applicable requirements of subsection (a) of this section.

"(2) The Administrator shall promulgate any water quality standard published in a proposed regulation not later than one hundred and ninety days after the date he publishes any such proposed standard, unless prior to such promulgation, such State has adopted a water quality standard which the Administrator determines to be in accordance with subsection (a) of this section.

Review. "(c)(1) The Governor of a State or the State water pollution control agency of such State shall from time to time (but at least once each three year period beginning with the date of enactment of the

154

Federal Water Pollution Control Act Amendments of 1972) hold public hearings for the purpose of reviewing applicable water quality standards and, as appropriate, modifying and adopting standards. Results of such review shall be made available to the Administrator.

"(2) Whenever the State revises or adopts a new standard, such revised or new standard shall be submitted to the Administrator. Such revised or new water quality standard shall consist of the designated uses of the navigable waters involved and the water quality criteria for such waters based upon such uses. Such standards shall be such as to protect the public health or welfare, enhance the quality of water and serve the purposes of this Act. Such standards shall be established taking into consideration their use and value for public water supplies, propagation of fish and wildlife, recreational purposes, and agricultural, industrial, and other purposes, and also taking into consideration their use and value for navigation.

"(3) If the Administrator, within sixty days after the date of submission of the revised or new standard, determines that such standard meets the requirements of this Act, such standard shall thereafter be the water quality standard for the applicable waters of that State. If the Administrator determines that any such revised or new standard is not consistent with the applicable requirements of this Act, he shall not later than the ninetieth day after the date of submission of such standard notify the State and specify the changes to meet such requirements. If such changes are not adopted by the State within ninety days after the date of notification, the Administrator shall promulgate such standard pursuant to paragraph (4) of this subsection.

"(4) The Administrator shall promptly prepare and publish proposed regulations setting forth a revised or new water quality standard for the navigable waters involved—

"(A) if a revised or new water quality standard submitted by such State under paragraph (3) of this subsection for such waters is determined by the Administrator not to be consistent with the applicable requirements of this Act, or

"(B) in any case where the Administrator determines that a revised or new standard is necessary to meet the requirements of this Act.

The Administrator shall promulgate any revised or new standard under this paragraph not later than ninety days after he publishes such proposed standards, unless prior to such promulgation, such State has adopted a revised or new water quality standard which the Administrator determines to be in accordance with this Act.

"(d) (1) (A) Each State shall identify those waters within its boundaries for which the effluent limitations required by section 301(b)(1) (A) and section 301(b)(1)(B) are not stringent enough to implement any water quality standard applicable to such waters. The State shall establish a priority ranking for such waters, taking into account the severity of the pollution and the uses to be made of such waters.

"(B) Each State shall identify those waters or parts thereof within its boundaries for which controls on thermal discharges under section 301 are not stringent enough to assure protection and propagation of a balanced indigenous population of shellfish, fish, and wildlife.

"(C) Each State shall establish for the waters identified in paragraph (1)(A) of this subsection, and in accordance with the priority ranking, the total maximum daily load, for those pollutants which the Administrator identifies under section 304(a)(2) as suitable for such calculation. Such load shall be established at a level necessary to implement the applicable water quality standards with seasonal variations and a margin of safety which takes into account any lack of

155

knowledge concerning the relationship between effluent limitations and water quality.

"(D) Each State shall estimate for the waters identified in paragraph (1)(B) of this subsection the total maximum daily thermal load required to assure protection and propagation of a balanced, indigenous population of shellfish, fish and wildlife. Such estimates shall take into account the normal water temperatures, flow rates, seasonal variations, existing sources of heat input, and the dissipative capacity of the identified waters or parts thereof. Such estimates shall include a calculation of the maximum heat input that can be made into each such part and shall include a margin of safety which takes into account any lack of knowledge concerning the development of thermal water quality criteria for such protection and propagation in the identified waters or parts thereof.

"(2) Each State shall submit to the Administrator from time to time, with the first such submission not later than one hundred and eighty days after the date of publication of the first identification of pollutants under section 304(a)(2)(D), for his approval the waters identified and the loads established under paragraphs (1)(A), (1)(B), (1)(C), and (1)(D) of this subsection. The Administrator shall either approve or disapprove such identification and load not later than thirty days after the date of submission. If the Administrator approves such identification and load, such State shall incorporate them into its current plan under subsection (e) of this section. If the Administrator disapproves such identification and load, he shall not later than thirty days after the date of such disapproval identify such waters in such State and establish such loads for such waters as he determines necessary to implement the water quality standards applicable to such waters and upon such identification and establishment the State shall incorporate them into its current plan under subsection (e) of this section.

"(3) For the specific purpose of developing information, each State shall identify all waters within its boundaries which it has not identified under paragraph (1)(A) and (1)(B) of this subsection and estimate for such waters the total maximum daily load with seasonal variations and margins of safety, for those pollutants which the Administrator identifies under section 304(a)(2) as suitable for such calculation and for thermal discharges, at a level that would assure protection and propagation of a balanced indigenous population of fish, shellfish and wildlife.

"(e)(1) Each State shall have a continuing planning process approved under paragraph (2) of this subsection which is consistent with this Act.

"(2) Each State shall submit not later than 120 days after the date of the enactment of the Water Pollution Control Amendments of 1972 to the Administrator for his approval a proposed continuing planning process which is consistent with this Act. Not later than thirty days after the date of submission of such a process the Administrator shall either approve or disapprove such process. The Administrator shall from time to time review each State's approved planning process for the purpose of insuring that such planning process is at all times consistent with this Act. The Administrator shall not approve any State permit program under title IV of this Act for any State which does not have an approved continuing planning process under this section.

"(3) The Administrator shall approve any continuing planning process submitted to him under this section which will result in plans for all navigable waters within such State, which include, but are not limited to, the following:

156

"(A) effluent limitations and schedules of compliance at least as stringent as those required by section 301(b)(1), section 301 (b)(2), section 306, and section 307. and at least as stringent as any requirements contained in any applicable water quality standard in effect under authority of this section;

"(B) the incorporation of all elements of any applicable area-wide waste management plans under section 208, and applicable basin plans under section 209 of this Act;

"(C) total maximum daily load for pollutants in accordance with subsection (d) of this section;

"(D) procedures for revision;

"(E) adequate authority for intergovernmental cooperation;

"(F) adequate implementation, including schedules of compliance, for revised or new water quality standards, under subsection (c) of this section;

"(G) controls over the disposition of all residual waste from any water treatment processing;

"(H) an inventory and ranking, in order of priority, of needs for construction of waste treatment works required to meet the applicable requirements of sections 301 and 302.

"(f) Nothing in this section shall be construed to affect any effluent limitation, or schedule of compliance required by any State to be implemented prior to the dates set forth in sections 301(b)(1) and 301 (b)(2) nor to preclude any State from requiring compliance with any effluent limitation or schedule of compliance at dates earlier than such dates.

"(g) Water quality standards relating to heat shall be consistent with the requirements of section 316 of this Act.   Heat standards.

"(h) For the purposes of this Act the term 'water quality standards' includes thermal water quality standards.   "Water quality standards."

"INFORMATION AND GUIDELINES

"Sec. 304. (a)(1) The Administrator, after consultation with appropriate Federal and State agencies and other interested persons, shall develop and publish, within one year after the date of enactment of this title (and from time to time thereafter revise) criteria for water quality accurately reflecting the latest scientific knowledge (A) on the kind and extent of all identifiable effects on health and welfare including, but not limited to, plankton, fish, shellfish, wildlife, plant life, shorelines, beaches, esthetics, and recreation which may be expected from the presence of pollutants in any body of water, including ground water; (B) on the concentration and dispersal of pollutants, or their byproducts, through biological, physical, and chemical processes; and (C) on the effects of pollutants on biological community diversity, productivity, and stability, including information on the factors affecting rates of eutrophication and rates of organic and inorganic sedimentation for varying types of receiving waters.   Water quality criteria publication.

"(2) The Administrator, after consultation with appropriate Federal and State agencies and other interested persons, shall develop and publish, within one year after the date of enactment of this title (and from time to time thereafter revise) information (A) on the factors necessary to restore and maintain the chemical, physical, and biological integrity of all navigable waters, ground waters, waters of the contiguous zone, and the oceans; (B) on the factors necessary for the protection and propagation of shellfish, fish, and wildlife for classes and categories of receiving waters and to allow recreational activities in and on the water; and (C) on the measurement and classification of water quality; and (D) for the purpose of section 303, on and the

157

identification of pollutants suitable for maximum daily load measurement correlated with the achievement of water quality objectives.

Publication in Federal Register. "(3) Such criteria and information and revisions thereof shall be issued to the States and shall be published in the Federal Register and otherwise made available to the public.

Effluent limitation guidelines, publication. "(b) For the purpose of adopting or revising effluent limitations under this Act the Administrator shall, after consultation with appropriate Federal and State agencies and other interested persons, publish within one year of enactment of this title, regulations, providing guidelines for effluent limitations, and, at least annually thereafter, revise, if appropriate, such regulations. Such regulations shall—

"(1)(A) identify, in terms of amounts of constituents and chemical, physical, and biological characteristics of pollutants, the degree of effluent reduction attainable through the application of the best practicable control technology currently available for classes and categories of point sources (other than publicly owned treatment works) ; and

"(B) specify factors to be taken into account in determining the control measures and practices to be applicable to point sources (other than publicly owned treatment works) within such categories or classes. Factors relating to the assessment of best practicable control technology currently available to comply with subsection (b)(1) of section 301 of this Act shall include consideration of the total cost of application of technology in relation to the effluent reduction benefits to be achieved from such application, and shall also take into account the age of equipment and facilities involved, the process employed, the engineering aspects of the application of various types of control techniques, process changes, non-water quality environmental impact (including energy requirements), and such other factors as the Administrator deems appropriate;

"(2)(A) identify, in terms of amounts of constituents and chemical, physical, and biological characteristics of pollutants, the degree of effluent reduction attainable through the application of the best control measures and practices achievable including treatment techniques, process and procedure innovations, operating methods, and other alternatives for classes and categories of point sources (other than publicly owned treatment works) ; and

"(B) specify factors to be taken into account in determining the best measures and practices available to comply with subsection (b)(2) of section 301 of this Act to be applicable to any point source (other than publicly owned treatment works) within such categories or classes. Factors relating to the assessment of best available technology shall take into account the age of equipment and facilities involved, the process employed, the engineering aspects of the application of various types of control techniques, process changes, the cost of achieving such effluent reduction, non-water quality environmental impact (including energy requirements), and such other factors as the Administrator deems appropriate; and

"(3) identify control measures and practices available to eliminate the discharge of pollutants from categories and classes of point sources, taking into account the cost of achieving such elimination of the discharge of pollutants.

Pollution discharges, elimination procedures information. "(c) The Administrator, after consultation, with appropriate Federal and State agencies and other interested persons, shall issue to the States and appropriate water pollution control agencies within 270 days after enactment of this title (and from time to time thereafter) information on the processes, procedures, or operating methods which result in the elimination or reduction of the discharge of pollutants to implement standards of performance under section 306 of this Act. Such information shall include technical and other data, including costs, as are available on alternative methods of elimination or reduction of the discharge of pollutants. Such information, and revisions thereof, shall be published in the Federal Register and otherwise shall be made available to the public.

Alternative waste treatment methods. Publication in Federal Register.

"(d)(1) The Administrator, after consultation with appropriate Federal and State agencies and other interested persons, shall publish within sixty days after enactment of this title (and from time to time thereafter) information, in terms of amounts of constituents and chemical, physical, and biological characteristics of pollutants, on the degree of effluent reduction attainable through the application of secondary treatment.

Secondary treatment information.

"(2) The Administrator, after consultation with appropriate Federal and State agencies and other interested persons, shall publish within nine months after the date of enactment of this title (and from time to time thereafter) information on alternative waste treatment management techniques and systems available to implement section 201 of this Act.

"(e) The Administrator, after consultation with appropriate Federal and State agencies and other interested persons, shall issue to appropriate Federal agencies, the States, water pollution control agencies, and agencies designated under section 208 of this Act, within one year after the effective date of this subsection (and from time to time thereafter) information including (1) guidelines for identifying and evaluating the nature and extent of nonpoint sources of pollutants, and (2) processes, procedures, and methods to control pollution resulting from—

"(A) agricultural and silvicultural activities, including runoff from fields and crop and forest lands;

"(B) mining activities, including runoff and siltation from new, currently operating, and abandoned surface and underground mines;

"(C) all construction activity, including runoff from the facilities resulting from such construction;

"(D) the disposal of pollutants in wells or in subsurface excavations;

"(E) salt water intrusion resulting from reductions of fresh water flow from any cause, including extraction of ground water, irrigation, obstruction, and diversion; and

"(F) changes in the movement, flow, or circulation of any navigable waters or ground waters, including changes caused by the construction of dams, levees, channels, causeways, or flow diversion facilities.

Such information and revisions thereof shall be published in the Federal Register and otherwise made available to the public.

Publication in Federal Register.

"(f)(1) For the purpose of assisting States in carrying out programs under section 402 of this Act, the Administrator shall publish, within one hundred and twenty days after the date of enactment of this title, and review at least annually thereafter and, if appropriate, revise guidelines for pretreatment of pollutants which he determines are not susceptible to treatment by publicly owned treatment works. Guidelines under this subsection shall be established to control and prevent the discharge into the navigable waters, the contiguous zone, or the ocean (either directly or through publicly owned treatment works) of any pollutant which interferes with, passes through, or otherwise is incompatible with such works.

Pretreatment standards guidelines, publication.

"(2) When publishing guidelines under this subsection, the Administrator shall designate the category or categories of treatment works to which the guidelines shall apply.

Test procedures, guidelines.

"(g) The Administrator shall, within one hundred and eighty days from the date of enactment of this title, promulgate guidelines establishing test procedures for the analysis of pollutants that shall include the factors which must be provided in any certification pursuant to section 401 of this Act or permit application pursuant to section 402 of this Act.

Monitoring, reporting, etc., guidelines.

"(h) The Administrator shall (1) within sixty days after the enactment of this title promulgate guidelines for the purpose of establishing uniform application forms and other minimum requirements for the acquisition of information from owners and operators of point-sources of discharge subject to any State program under section 402 of this Act, and (2) within sixty days from the date of enactment of this title promulgate guidelines establishing the minimum procedural and other elements of any State program under section 402 of this Act which shall include:

"(A) monitoring requirements;

"(B) reporting requirements (including procedures to make information available to the public);

"(C) enforcement provisions; and

"(D) funding, personnel qualifications, and manpower requirements (including a requirement that no board or body which approves permit applications or portions thereof shall include, as a member, any person who receives, or has during the previous two years received, a significant portion of his income directly or indirectly from permit holders or applicants for a permit).

"(i) The Administrator shall, within 270 days after the effective date of this subsection (and from time to time thereafter), issue such information on methods, procedures, and processes as may be appropriate to restore and enhance the quality of the Nation's publicly owned fresh water lakes.

"(j)(1) The Administrator shall, within six months from the date of enactment of this title, enter into agreements with the Secretary of Agriculture, the Secretary of the Army, and the Secretary of the Interior to provide for the maximum utilization of the appropriate programs authorized under other Federal law to be carried out by such Secretaries for the purpose of achieving and maintaining water quality through appropriate implementation of plans approved under section 208 of this Act.

Transfer of funds.

"(2) The Administrator, pursuant to any agreement under paragraph (1) of this subsection is authorized to transfer to the Secretary of Agriculture, the Secretary of the Army, or the Secretary of the Interior any funds appropriated under paragraph (3) of this subsection to supplement any funds otherwise appropriated to carry out appropriate programs authorized to be carried out by such Secretaries.

Appropriation.

"(3) There is authorized to be appropriated to carry out the provisions of this subsection, $100,000,000 per fiscal year for the fiscal year ending June 30, 1973, and the fiscal year ending June 30, 1974.

"WATER QUALITY INVENTORY

Report to Congress.

"SEC. 305. (a) The Administrator, in cooperation with the States and with the assistance of appropriate Federal agencies, shall prepare a report to be submitted to the Congress on or before January 1, 1974, which shall—

"(1) describe the specific quality, during 1973, with appropriate supplemental descriptions as shall be required to take into account seasonal, tidal, and other variations, of all navigable waters and the waters of the contiguous zone;

"(2) include an inventory of all point sources of discharge (based on a qualitative and quantitative analysis of discharges) of pollutants, into all navigable waters and the waters of the contiguous zone; and

"(3) identify specifically those navigable waters, the quality of which—

"(A) is adequate to provide for the protection and propagation of a balanced population of shellfish, fish, and wildlife and allow recreational activities in and on the water;

"(B) can reasonably be expected to attain such level by 1977 or 1983; and

"(C) can reasonably be expected to attain such level by any later date.

"(b) (1) Each State shall prepare and submit to the Administrator by January 1, 1975, and shall bring up to date each year thereafter, a report which shall include—

State reports.

"(A) a description of the water quality of all navigable waters in such State during the preceding year, with appropriate supplemental descriptions as shall be required to take into account seasonal, tidal, and other variations, correlated with the quality of water required by the objective of this Act (as identified by the Administrator pursuant to criteria published under section 304(a) of this Act) and the water quality described in subparagraph (B) of this paragraph;

"(B) an analysis of the extent to which all navigable waters of such State provide for the protection and propagation of a balanced population of shellfish, fish, and wildlife, and allow recreational activities in and on the water;

"(C) an analysis of the extent to which the elimination of the discharge of pollutants and a level of water quality which provides for the protection and propagation of a balanced population of shellfish, fish, and wildlife and allows recreational activities in and on the water, have been or will be achieved by the requirements of this Act, together with recommendations as to additional action necessary to achieve such objectives and for what waters such additional action is necessary;

"(D) an estimate of (i) the environmental impact, (ii) the economic and social costs necessary to achieve the objective of this Act in such State, (iii) the economic and social benefits of such achievement, and (iv) an estimate of the date of such achievement; and

"(E) a description of the nature and extent of nonpoint sources of pollutants, and recommendations as to the programs which must be undertaken to control each category of such sources, including an estimate of the costs of implementing such programs.

"(2) The Administrator shall transmit such State reports, together with an analysis thereof, to Congress on or before October 1, 1975, and annually thereafter.

Transmittal to Congress.

"NATIONAL STANDARDS OF PERFORMANCE

"Sec. 306. (a) For purposes of this section:

Definitions.

"(1) The term 'standard of performance' means a standard for the control of the discharge of pollutants which reflects the greatest degree of effluent reduction which the Administrator determines to be achievable through application of the best available demonstrated control technology, processes, operating methods, or other alterna-

161

tives, including, where practicable, a standard permitting no discharge of pollutants.

"(2) The term 'new source' means any source, the construction of which is commenced after the publication of proposed regulations prescribing a standard of performance under this section which will be applicable to such source, if such standard is thereafter promulgated in accordance with this section.

"(3) The term 'source' means any building, structure, facility, or installation from which there is or may be the discharge of pollutants.

"(4) The term 'owner or operator' means any person who owns, leases, operates, controls, or supervises a source.

"(5) The term 'construction' means any placement, assembly, or installation of facilities or equipment (including contractual obligations to purchase such facilities or equipment) at the premises where such equipment will be used, including preparation work at such premises.

Sources, category list, publication. "(b) (1) (A) The Administrator shall, within ninety days after the date of enactment of this title publish (and from time to time thereafter shall revise) a list of categories of sources, which shall, at the minimum, include:

> "pulp and paper mills;
> "paperboard, builders paper and board mills;
> "meat product and rendering processing;
> "dairy product processing;
> "grain mills;
> "canned and preserved fruits and vegetables processing;
> "canned and preserved seafood processing;
> "sugar processing;
> "textile mills;
> "cement manufacturing;
> "feedlots;
> "electroplating;
> "organic chemicals manufacturing;
> "inorganic chemicals manufacturing;
> "plastic and synthetic materials manufacturing;
> "soap and detergent manufacturing;
> "fertilizer manufacturing;
> "petroleum refining;
> "iron and steel manufacturing;
> "nonferrous metals manufacturing;
> "phosphate manufacturing;
> "steam electric powerplants;
> "ferroalloy manufacturing;
> "leather tanning and finishing;
> "glass and asbestos manufacturing;
> "rubber processing; and
> "timber products processing.

Federal standards of performance, publication of regulations. "(B) As soon as practicable, but in no case more than one year, after a category of sources is included in a list under subparagraph (A) of this paragraph, the Administrator shall propose and publish regulations establishing Federal standards of performance for new sources within such category. The Administrator shall afford interested persons an opportunity for written comment on such proposed regulations. After considering such comments, he shall promulgate, within one hundred and twenty days after publication of such proposed regulations, such standards with such adjustments as he deems appropriate. The Administrator shall, from time to time, as technology and alternatives change, revise such standards following the procedure required by this subsection for promulgation of such standards. Standards of performance, or revisions thereof, shall become effective upon promulgation. In establishing or revising Federal standards of performance for new sources under this section, the Administrator shall take into consideration the cost of achieving such effluent reduction, and any non-water quality environmental impact and energy requirements.

"(2) The Administrator may distinguish among classes, types, and sizes within categories of new sources for the purpose of establishing such standards and shall consider the type of process employed (including whether batch or continuous).

"(3) The provisions of this section shall apply to any new source owned or operated by the United States.

"(c) Each State may develop and submit to the Administrator a procedure under State law for applying and enforcing standards of performance for new sources located in such State. If the Administrator finds that the procedure and the law of any State require the application and enforcement of standards of performance to at least the same extent as required by this section, such State is authorized to apply and enforce such standards of performance (except with respect to new sources owned or operated by the United States). Standards of performance, State enforcement procedure.

"(d) Notwithstanding any other provision of this Act, any point source the construction of which is commenced after the date of enactment of the Federal Water Pollution Control Act Amendments of 1972 and which is so constructed as to meet all applicable standards of performance shall not be subject to any more stringent standard of performance during a ten-year period beginning on the date of completion of such construction or during the period of depreciation or amortization of such facility for the purposes of section 167 or 169 (or both) of the Internal Revenue Code of 1954, whichever period ends first. 68A Stat. 51; 85 Stat. 508. 83 Stat. 667. 26 USC 167, 169.

"(e) After the effective date of standards of performance promulgated under this section, it shall be unlawful for any owner or operator of any new source to operate such source in violation of any standard of performance applicable to such source.

"TOXIC AND PRETREATMENT EFFLUENT STANDARDS

"Sec. 307. (a) (1) The Administrator shall, within ninety days after the date of enactment of this title, publish (and from time to time thereafter revise) a list which includes any toxic pollutant or combination of such pollutants for which an effluent standard (which may include a prohibition of the discharge of such pollutants or combination of such pollutants) will be established under this section. The Administrator in publishing such list shall take into account the toxicity of the pollutant, its persistence, degradability, the usual or potential presence of the affected organisms in any waters, the importance of the affected organisms and the nature and extent of the effect of the toxic pollutant on such organisms.

"(2) Within one hundred and eighty days after the date of publication of any list, or revision thereof, containing toxic pollutants or combination of pollutants under paragraph (1) of this subsection, the Administrator, in accordance with section 553 of title 5 of the United States Code, shall publish a proposed effluent standard (or a prohibition) for such pollutant or combination of pollutants which shall take into account the toxicity of the pollutant, its persistence, degradability, Proposed effluent standard, publication. 80 Stat. 383.

the usual or potential presence of the affected organisms in any waters, the importance of the affected organisms and the nature and extent of the effect of the toxic pollutant on such organisms, and he shall publish a notice for a public hearing on such proposed standard to be held within thirty days. As soon as possible after such hearing, but not later than six months after publication of the proposed effluent standard (or prohibition), unless the Administrator finds, on the record, that a modification of such proposed standard (or prohibition) is justified based upon a preponderance of evidence adduced at such hearings, such standard (or prohibition) shall be promulgated.

"(3) If after a public hearing the Administrator finds that a modification of such proposed standard (or prohibition) is justified, a revised effluent standard (or prohibition) for such pollutant or combination of pollutants shall be promulgated immediately. Such standard (or prohibition) shall be reviewed and, if appropriate, revised at least every three years.

"(4) Any effluent standard promulgated under this section shall be at that level which the Administrator determines provides an ample margin of safety.

"(5) When proposing or promulgating any effluent standard (or prohibition) under this section, the Administrator shall designate the category or categories of sources to which the effluent standard (or prohibition) shall apply. Any disposal of dredged material may be included in such a category of sources after consultation with the Secretary of the Army.

"(6) Any effluent standard (or prohibition) established pursuant to this section shall take effect on such date or dates as specified in the order promulgating such standard, but in no case more than one year from the date of such promulgation.

"(7) Prior to publishing any regulations pursuant to this section the Administrator shall, to the maximum extent practicable within the time provided, consult with appropriate advisory committees, States, independent experts, and Federal departments and agencies.

"(b)(1) The Administrator shall, within one hundred and eighty days after the date of enactment of this title and from time to time thereafter, publish proposed regulations establishing pretreatment standards for introduction of pollutants into treatment works (as defined in section 212 of this Act) which are publicly owned for those pollutants which are determined not to be susceptible to treatment by such treatment works or which would interfere with the operation of such treatment works. Not later than ninety days after such publication, and after opportunity for public hearing, the Administrator shall promulgate such pretreatment standards. Pretreatment standards under this subsection shall specify a time for compliance not to exceed three years from the date of promulgation and shall be established to prevent the discharge of any pollutant through treatment works (as defined in section 212 of this Act) which are publicly owned, which pollutant interferes with, passes through, or otherwise is incompatible with such works.

"(2) The Administrator shall, from time to time, as control technology, processes, operating methods, or other alternatives change, revise such standards following the procedure established by this subsection for promulgation of such standards.

"(3) When proposing or promulgating any pretreatment standard under this section, the Administrator shall designate the category or categories of sources to which such standard shall apply.

"(4) Nothing in this subsection shall affect any pretreatment requirement established by any State or local law not in conflict with any pretreatment standard established under this subsection.

164

"(c) In order to insure that any source introducing pollutants into a publicly owned treatment works, which source would be a new source subject to section 306 if it were to discharge pollutants, will not cause a violation of the effluent limitations established for any such treatment works, the Administrator shall promulgate pretreatment standards for the category of such sources simultaneously with the promulgation of standards of performance under section 306 for the equivalent category of new sources. Such pretreatment standards shall prevent the discharge of any pollutant into such treatment works, which pollutant may interfere with, pass through, or otherwise be incompatible with such works.

"(d) After the effective date of any effluent standard or prohibition or pretreatment standard promulgated under this section, it shall be unlawful for any owner or operator of any source to operate any source in violation of any such effluent standard or prohibition or pretreatment standard.

"INSPECTIONS, MONITORING AND ENTRY

"SEC. 308. (a) Whenever required to carry out the objective of this Act, including but not limited to (1) developing or assisting in the development of any effluent limitation, or other limitation, prohibition, or effluent standard, pretreatment standard, or standard of performance under this Act; (2) determining whether any person is in violation of any such effluent limitation, or other limitation, prohibition or effluent standard, pretreatment standard, or standard of performance; (3) any requirement established under this section; or (4) carrying out sections 305, 311, 402, and 504 of this Act—

"(A) the Administrator shall require the owner or operator of any point source to (i) establish and maintain such records, (ii) make such reports, (iii) install, use, and maintain such monitoring equipment or methods (including where appropriate, biological monitoring methods), (iv) sample such effluents (in accordance with such methods, at such locations, at such intervals, and in such manner as the Administrator shall prescribe), and (v) provide such other information as he may reasonably require; and

Recordkeeping; reports.

"(B) the Administrator or his authorized representative, upon presentation of his credentials—

"(i) shall have a right of entry to, upon, or through any premises in which an effluent source is located or in which any records required to be maintained under clause (A) of this subsection are located, and

"(ii) may at reasonable times have access to and copy any records, inspect any monitoring equipment or method required under clause (A), and sample any effluents which the owner or operator of such source is required to sample under such clause.

"(b) Any records, reports, or information obtained under this section (1) shall, in the case of effluent data, be related to any applicable effluent limitations, toxic, pretreatment, or new source performance standards, and (2) shall be available to the public, except that upon a showing satisfactory to the Administrator by any person that records, reports, or information, or particular part thereof (other than effluent data), to which the Administrator has access under this section, if made public would divulge methods or processes entitled to protection as trade secrets of such person, the Administrator shall consider such record, report, or information, or particular portion thereof confidential in accordance with the purposes of section 1905 of title 18 of the United States Code, except that such record, report, or information may be disclosed to other officers, employees, or authorized representatives of the United States concerned with carrying out this Act or when relevant in any proceeding under this Act.

"(c) Each State may develop and submit to the Administrator procedures under State law for inspection, monitoring, and entry with respect to point sources located in such State. If the Administrator finds that the procedures and the law of any State relating to inspection, monitoring, and entry are applicable to at least the same extent as those required by this section, such State is authorized to apply and enforce its procedures for inspection, monitoring, and entry with respect to point sources located in such State (except with respect to point sources owned or operated by the United States).

"FEDERAL ENFORCEMENT

"SEC. 309. (a) (1) Whenever, on the basis of any information available to him, the Administrator finds that any person is in violation of any condition or limitation which implements section 301, 302, 306, 307, or 308 of this Act in a permit issued by a State under an approved permit program under section 402 of this Act, he shall proceed under his authority in paragraph (3) of this subsection or he shall notify the person in alleged violation and such State of such finding. If beyond the thirtieth day after the Administrator's notification the State has not commenced appropriate enforcement action, the Administrator shall issue an order requiring such person to comply with such condition or limitation or shall bring a civil action in accordance with subsection (b) of this section.

"(2) Whenever, on the basis of information available to him, the Administrator finds that violations of permit conditions or limitations as set forth in paragraph (1) of this subsection are so widespread that such violations appear to result from a failure of the State to enforce such permit conditions or limitations effectively, he shall so notify the State. If the Administrator finds such failure extends beyond the thirtieth day after such notice, he shall give public notice of such finding. During the period beginning with such public notice and ending when such State satisfies the Administrator that it will enforce such conditions and limitations (hereafter referred to in this section as the period of 'federally assumed enforcement'), the Administrator shall enforce any permit condition or limitation with respect to any person—

"(A) by issuing an order to comply with such condition or limitation, or

"(B) by bringing a civil action under subsection (b) of this section.

"(3) Whenever on the basis of any information available to him the Administrator finds that any person is in violation of section 301, 302, 306, 307, or 308 of this Act, or is in violation of any permit condition or limitation implementing any of such sections in a permit issued under section 402 of this Act by him or by a State, he shall issue an order requiring such person to comply with such section or requirement, or he shall bring a civil action in accordance with subsection (b) of this section.

"(4) A copy of any order issued under this subsection shall be sent immediately by the Administrator to the State in which the violation occurs and other affected States. Any order issued under this subsection shall be by personal service and shall state with reasonable specificity the nature of the violation, specify a time for compliance, not to exceed thirty days, which the Administrator determines is reasonable, taking into account the seriousness of the violation and any good faith efforts

166

to comply with applicable requirements. In any case in which an order under this subsection (or notice to a violator under paragraph (1) of this subsection) is issued to a corporation, a copy of such order (or notice) shall be served on any appropriate corporate officers. An order issued under this subsection relating to a violation of section 308 of this Act shall not take effect until the person to whom it is issued has had an opportunity to confer with the Administrator concerning the alleged violation.

"(b) The Administrator is authorized to commence a civil action for appropriate relief, including a permanent or temporary injunction, for any violation for which he is authorized to issue a compliance order under subsection (a) of this section. Any action under this subsection may be brought in the district court of the United States for the district in which the defendant is located or resides or is doing business, and such court shall have jurisdiction to restrain such violation and to require compliance. Notice of the commencement of such action shall be given immediately to the appropriate State.

"(c) (1) Any person who willfully or negligently violates section 301, 302, 306, 307, or 308 of this Act, or any permit condition or limitation implementing any of such sections in a permit issued under section 402 of this Act by the Administrator or by a State, shall be punished by a fine of not less than $2,500 nor more than $25,000 per day of violation, or by imprisonment for not more than one year, or by both. If the conviction is for a violation committed after a first conviction of such person under this paragraph, punishment shall be by a fine of not more than $50,000 per day of violation, or by imprisonment for not more than two years, or by both. *Penalties.*

"(2) Any person who knowingly makes any false statement, representation, or certification in any application, record, report, plan, or other document filed or required to be maintained under this Act or who falsifies, tampers with, or knowingly renders inaccurate any monitoring device or method required to be maintained under this Act, shall upon conviction, be punished by a fine of not more than $10,000, or by imprisonment for not more than six months, or by both.

"(3) For the purposes of this subsection, the term 'person' shall mean, in addition to the definition contained in section 502(5) of this Act, any responsible corporate officer. *"Person."*

"(d) Any person who violates section 301, 302, 306, 307, or 308 of this Act, or any permit condition or limitation implementing any of such sections in a permit issued under section 402 of this Act by the Administrator, or by a State, and any person who violates any order issued by the Administrator under subsection (a) of this section, shall be subject to a civil penalty not to exceed $10,000 per day of such violation.

"(e) Whenever a municipality is a party to a civil action brought by the United States under this section, the State in which such municipality is located shall be joined as a party. Such State shall be liable for payment of any judgment, or any expenses incurred as a result of complying with any judgment, entered against the municipality in such action to the extent that the laws of that State prevent the municipality from raising revenues needed to comply with such judgment.

"INTERNATIONAL POLLUTION ABATEMENT

"SEC. 310. (a) Whenever the Administrator, upon receipts of reports, surveys, or studies from any duly constituted international agency, has reason to believe that pollution is occurring which endangers the health or welfare of persons in a foreign country, and the Secretary of State requests him to abate such pollution, he shall give

167

formal notification thereof to the State water pollution control agency of the State or States in which such discharge or discharges originate and to the appropriate interstate agency, if any. He shall also promptly call such a hearing, if he believes that such pollution is occurring in sufficient quantity to warrant such action, and if such foreign country has given the United States essentially the same rights with respect to the prevention and control of pollution occurring in that country as is given that country by this subsection. The Administrator, through the Secretary of State, shall invite the foreign country which may be adversely affected by the pollution to attend and participate in the hearing, and the representative of such country shall, for the purpose of the hearing and any further proceeding resulting from such hearing, have all the rights of a State water pollution control agency. Nothing in this subsection shall be construed to modify, amend, repeal, or otherwise affect the provisions of the 1909 Boundary Waters Treaty between Canada and the United States or the Water Utilization Treaty of 1944 between Mexico and the United States (59 Stat. 1219), relative to the control and abatement of pollution in waters covered by those treaties.

Hearing.

36 Stat. 2448.

"(b) The calling of a hearing under this section shall not be construed by the courts, the Administrator, or any person as limiting, modifying, or otherwise affecting the functions and responsibilities of the Administrator under this section to establish and enforce water quality requirements under this Act.

"(c) The Administrator shall publish in the Federal Register a notice of a public hearing before a hearing board of five or more persons appointed by the Administrator. A majority of the members of the board and the chairman who shall be designated by the Administrator shall not be officers or employees of Federal, State, or local governments. On the basis of the evidence presented at such hearing, the board shall within sixty days after completion of the hearing make findings of fact as to whether or not such pollution is occurring and shall thereupon by decision, incorporating its findings therein, make such recommendations to abate the pollution as may be appropriate and shall transmit such decision and the record of the hearings to the Administrator. All such decisions shall be public. Upon receipt of such decision, the Administrator shall promptly implement the board's decision in accordance with the provisions of this Act.

Notice, publication in Federal Register.

"(d) In connection with any hearing called under this subsection, the board is authorized to require any person whose alleged activities result in discharges causing or contributing to pollution to file with it in such forms as it may prescribe, a report based on existing data, furnishing such information as may reasonably be required as to the character, kind, and quantity of such discharges and the use of facilities or other means to prevent or reduce such discharges by the person filing such a report. Such report shall be made under oath or otherwise, as the board may prescribe, and shall be filed with the board within such reasonable period as it may prescribe, unless additional time is granted by it. Upon a showing satisfactory to the board by the person filing such report that such report or portion thereof (other than effluent data), to which the Administrator has access under this section, if made public would divulge trade secrets or secret processes of such person, the board shall consider such report or portion thereof confidential for the purposes of section 1905 of title 18 of the United States Code. If any person required to file any report under this paragraph shall fail to do so within the time fixed by the board for filing the same, and such failure shall continue for thirty days after notice of such default, such person shall forfeit to the United States the sum

Report.

62 Stat. 791.
Penalty.

168

of $1,000 for each and every day of the continuance of such failure, which forfeiture shall be payable into the Treasury of the United States, and shall be recoverable in a civil suit in the name of the United States in the district court of the United States where such person has his principal office or in any district in which he does business. The Administrator may upon application therefor remit or mitigate any forfeiture provided for under this subsection.

"(e) Board members, other than officers or employees of Federal, State, or local governments, shall be for each day (including travel-time) during which they are performing board business, entitled to receive compensation at a rate fixed by the Administrator but not in excess of the maximum rate of pay for grade GS–18, as provided in the General Schedule under section 5332 of title 5 of the United States Code, and shall, notwithstanding the limitations of sections 5703 and 5704 of title 5 of the United States Code, be fully reimbursed for travel, subsistence, and related expenses.

80 Stat. 499;
83 Stat. 190.

"(f) When any such recommendation adopted by the Administrator involves the institution of enforcement proceedings against any person to obtain the abatement of pollution subject to such recommendation, the Administrator shall institute such proceedings if he believes that the evidence warrants such proceedings. The district court of the United States shall consider and determine de novo all relevant issues, but shall receive in evidence the record of the proceedings before the conference or hearing board. The court shall have jurisdiction to enter such judgment and orders enforcing such judgment as it deems appropriate or to remand such proceedings to the Administrator for such further action as it may direct.

"OIL AND HAZARDOUS SUBSTANCE LIABILITY

"SEC. 311. (a) For the purpose of this section, the term—   Definitions.
"(1) 'oil' means oil of any kind or in any form, including, but not limited to, petroleum, fuel oil, sludge, oil refuse, and oil mixed with wastes other than dredged spoil;
"(2) 'discharge' includes, but is not limited to, any spilling, leaking, pumping, pouring, emitting, emptying or dumping;
"(3) 'vessel' means every description of watercraft or other artificial contrivance used, or capable of being used, as a means of transportation on water other than a public vessel;
"(4) 'public vessel' means a vessel owned or bareboat-chartered and operated by the United States, or by a State or political subdivision thereof, or by a foreign nation, except when such vessel is engaged in commerce;
"(5) 'United States' means the States, the District of Columbia, the Commonwealth of Puerto Rico, the Canal Zone, Guam, American Samoa, the Virgin Islands, and the Trust Territory of the Pacific Islands;
"(6) 'owner or operator' means (A) in the case of a vessel, any person owning, operating, or chartering by demise, such vessel, and (B) in the case of an onshore facility, and an offshore facility, any person owning or operating such onshore facility or offshore facility, and (C) in the case of any abandoned offshore facility, the person who owned or operated such facility immediately prior to such abandonment;
"(7) 'person' includes an individual, firm, corporation, association, and a partnership;
"(8) 'remove' or 'removal' refers to removal of the oil or hazardous substances from the water and shorelines or the taking of

such other actions as may be necessary to minimize or mitigate damage to the public health or welfare, including, but not limited to, fish, shellfish, wildlife, and public and private property, shorelines, and beaches;

"(9) 'contiguous zone' means the entire zone established or to be established by the United States under article 24 of the Convention on the Territorial Sea and the Contiguous Zone;

15 UST 1606.

"(10) 'onshore facility' means any facility (including, but not limited to, motor vehicles and rolling stock) of any kind located in, on, or under, any land within the United States other than submerged land;

"(11) 'offshore facility' means any facility of any kind located in, on, or under, any of the navigable waters of the United States other than a vessel or a public vessel;

"(12) 'act of God' means an act occasioned by an unanticipated grave natural disaster;

"(13) 'barrel' means 42 United States gallons at 60 degrees Fahrenheit;

"(14) 'hazardous substance' means any substance designated pursuant to subsection (b)(2) of this section.

Prohibition.

"(b)(1) The Congress hereby declares that it is the policy of the United States that there should be no discharges of oil or hazardous substances into or upon the navigable waters of the United States, adjoining shorelines, or into or upon the waters of the contiguous zone.

Regulations.

"(2)(A) The Administrator shall develop, promulgate, and revise as may be appropriate, regulations designating as hazardous substances, other than oil as defined in this section, such elements and compounds which, when discharged in any quantity into or upon the navigable waters of the United States or adjoining shorelines or the waters of the contiguous zone, present an imminent and substantial danger to the public health or welfare, including, but not limited to, fish, shellfish, wildlife, shorelines, and beaches.

Hazardous substances, removability determination.

"(B)(i) The Administrator shall include in any designation under subparagraph (A) of this subsection a determination whether any such designated hazardous substance can actually be removed.

"(ii) The owner or operator of any vessel, onshore facility, or offshore facility from which there is discharged during the two-year period beginning on the date of enactment of the Federal Water Pollution Control Act Amendments of 1972, any hazardous substance determined not removable under clause (i) of this subparagraph shall be liable, subject to the defenses to liability provided under subsection (f) of this section, as appropriate, to the United States for a civil penalty per discharge established by the Administrator based on toxicity, degradability, and dispersal characteristics of such substance, in an amount not to exceed $50,000, except that where the United States can show that such discharge was a result of willful negligence or willful misconduct within the privity and knowledge of the owner, such owner or operator shall be liable to the United States for a civil penalty in such amount as the Administrator shall establish, based upon the toxicity, degradability, and dispersal characteristics of such substance.

"(iii) After the expiration of the two-year period referred to in clause (ii) of this subparagraph, the owner or operator of any vessel, onshore facility, or offshore facility, from which there is discharged any hazardous substance determined not removable under clause (i) of this subparagraph shall be liable, subject to the defenses to liability provided in subsection (f) of this section, to the United States for either one or the other of the following penalties, the determination of which shall be in the discretion of the Administrator:

170

"(aa) a penalty in such amount as the Administrator shall establish, based on the toxicity, degradability, and dispersal characteristics of the substance, but not less than $500 nor more than $5,000; or

"(bb) a penalty determined by the number of units discharged multiplied by the amount established for such unit under clause (iv) of this subparagraph, but such penalty shall not be more than $5,000,000 in the case of a discharge from a vessel and $500,000 in the case of a discharge from an onshore or offshore facility.

"(iv) The Administrator shall establish by regulation, for each hazardous substance designated under subparagraph (A) of this paragraph, and within 180 days of the date of such designation, a unit of measurement based upon the usual trade practice and, for the purpose of determining the penalty under clause (iii)(bb) of this subparagraph, shall establish for each such unit a fixed monetary amount which shall be not less than $100 nor more than $1,000 per unit. He shall establish such fixed amount based on the toxicity, degradability, and dispersal characteristics of the substance.

"(3) The discharge of oil or hazardous substances into or upon the navigable waters of the United States, adjoining shorelines, or into or upon the waters of the contiguous zone in harmful quantities as determined by the President under paragraph (4) of this subsection, is prohibited, except (A) in the case of such discharges of oil into the waters of the contiguous zone, where permitted under article IV of the International Convention for the Prevention of Pollution of the Sea by Oil, 1954, as amended, and (B) where permitted in quantities and 12 UST 2989. at times and locations or under such circumstances or conditions as the President may, by regulation, determine not to be harmful. Any regulations issued under this subsection shall be consistent with maritime safety and with marine and navigation laws and regulations and applicable water quality standards.

"(4) The President shall by regulation, to be issued as soon as possible after the date of enactment of this paragraph, determine for the purposes of this section, those quantities of oil and any hazardous substance the discharge of which, at such times, locations, circumstances, and conditions, will be harmful to the public health or welfare of the United States, including, but not limited to, fish, shellfish, wildlife, and public and private property, shorelines, and beaches except that in the case of the discharge of oil into or upon the waters of the contiguous zone, only those discharges which threaten the fishery resources of the contiguous zone or threaten to pollute or contribute to the pollution of the territory or the territorial sea of the United States may be determined to be harmful. **Discharges, harmful quantities, determination.**

"(5) Any person in charge of a vessel or of an onshore facility or an offshore facility shall, as soon as he has knowledge of any discharge of oil or a hazardous substance from such vessel or facility in violation of paragraph (3) of this subsection, immediately notify the appropriate agency of the United States Government of such discharge. Any such person who fails to notify immediately such agency of such discharge shall, upon conviction, be fined not more than $10,000, or imprisoned for not more than one year, or both. Notification received pursuant to this paragraph or information obtained by the exploitation of such notification shall not be used against any such person in any criminal case, except a prosecution for perjury or for giving a false statement. **Violation notification.** **Penalty.**

"(6) Any owner or operator of any vessel, onshore facility, or offshore facility from which oil or a hazardous substance is discharged in violation of paragraph (3) of this subsection shall be assessed a civil penalty by the Secretary of the department in which the Coast Guard is operating of not more than $5,000 for each offense. No penalty shall be assessed unless the owner or operator charged shall have been given notice and opportunity for a hearing on such charge. Each violation is a separate offense. Any such civil penalty may be compromised by such Secretary. In determining the amount of the penalty, or the amount agreed upon in compromise, the appropriateness of such penalty to the size of the business of the owner or operator charged, the effect on the owner or operator's ability to continue in business, and the gravity of the violation, shall be considered by such Secretary. The Secretary of the Treasury shall withhold at the request of such Secretary the clearance required by section 4197 of the Revised Statutes of the United States, as amended (46 U.S.C. 91), of any vessel the owner or operator of which is subject to the foregoing penalty. Clearance may be granted in such cases upon the filing of a bond or other surety satisfactory to such Secretary.

Discharge into
U.S. navigable
waters, removal.

"(c)(1) Whenever any oil or a hazardous substance is discharged, into or upon the navigable waters of the United States, adjoining shorelines, or into or upon the waters of the contiguous zone, the President is authorized to act to remove or arrange for the removal of such oil or substance at any time, unless he determines such removal will be done properly by the owner or operator of the vessel, onshore facility, or offshore facility from which the discharge occurs.

National
Contingency
Plan.

"(2) Within sixty days after the effective date of this section, the President shall prepare and publish a National Contingency Plan for removal of oil and hazardous substances, pursuant to this subsection. Such National Contingency Plan shall provide for efficient, coordinated, and effective action to minimize damage from oil and hazardous substance discharges, including containment, dispersal, and removal of oil and hazardous substances, and shall include, but not be limited to—

"(A) assignment of duties and responsibilities among Federal departments and agencies in coordination with State and local agencies, including, but not limited to, water pollution control, conservation, and port authorities;

"(B) identification, procurement, maintenance, and storage of equipment and supplies;

"(C) establishment or designation of a strike force consisting of personnel who shall be trained, prepared, and available to provide necessary services to carry out the Plan, including the establishment at major ports, to be determined by the President, of emergency task forces of trained personnel, adequate oil and hazardous substance pollution control equipment and material, and a detailed oil and hazardous substance pollution prevention and removal plan;

"(D) a system of surveillance and notice designed to insure earliest possible notice of discharges of oil and hazardous substances to the appropriate Federal agency;

"(E) establishment of a national center to provide coordination and direction for operations in carrying out the Plan;

"(F) procedures and techniques to be employed in identifying, containing, dispersing, and removing oil and hazardous substances;

"(G) a schedule, prepared in cooperation with the States, identifying (i) dispersants and other chemicals, if any, that may be used in carrying out the Plan, (ii) the waters in which such dispersants and chemicals may be used, and (iii) the quantities of such dispersant or chemical which can be used safely in such waters, which schedule shall provide in the case of any dispersant, chemical, or waters not specifically identified in such schedule that the President, or his delegate, may, on a case-by-case basis, iden-

172

tify the dispersants and other chemicals which may be used, the waters in which they may be used, and the quantities which can be used safely in such waters; and

"(H) a system whereby the State or States affected by a discharge of oil or hazardous substance may act where necessary to remove such discharge and such State or States may be reimbursed from the fund established under subsection (k) of this section for the reasonable costs incurred in such removal.

The President may, from time to time, as he deems advisable revise or otherwise amend the National Contingency Plan. After publication of the National Contingency Plan, the removal of oil and hazardous substances and actions to minimize damage from oil and hazardous substance discharges shall, to the greatest extent possible, be in accordance with the National Contingency Plan.

"(d) Whenever a marine disaster in or upon the navigable waters of the United States has created a substantial threat of a pollution hazard to the public health or welfare of the United States, including, but not limited to, fish, shellfish, and wildlife and the public and private shorelines and beaches of the United States, because of a discharge, or an imminent discharge, of large quantities of oil, or of a hazardous substance from a vessel the United States may (A) coordinate and direct all public and private efforts directed at the removal or elimination of such threat; and (B) summarily remove, and, if necessary, destroy such vessel by whatever means are available without regard to any provisions of law governing the employment of personnel or the expenditure of appropriated funds. Any expense incurred under this subsection shall be a cost incurred by the United States Government for the purposes of subsection (f) in the removal of oil or hazardous substance. *Maritime disaster discharge.*

"(e) In addition to any other action taken by a State or local government, when the President determines there is an imminent and substantial threat to the public health or welfare of the United States, including, but not limited to, fish, shellfish, and wildlife and public and private property, shorelines, and beaches within the United States, because of an actual or threatened discharge of oil or hazardous substance into or upon the navigable waters of the United States from an onshore or offshore facility, the President may require the United States attorney of the district in which the threat occurs to secure such relief as may be necessary to abate such threat, and the district courts of the United States shall have jurisdiction to grant such relief as the public interest and the equities of the case may require. *Relief.* ... *Jurisdiction.*

"(f)(1) Except where an owner or operator can prove that a discharge was caused solely by (A) an act of God, (B) an act of war, (C) negligence on the part of the United States Government, or (D) an act or omission of a third party without regard to whether any such act or omission was or was not negligent, or any combination of the foregoing clauses, such owner or operator of any vessel from which oil or a hazardous substance is discharged in violation of subsection (b)(2) of this section shall, not withstanding any other provision of law, be liable to the United States Government for the actual costs incurred under subsection (c) for the removal of such oil or substance by the United States Government in an amount not to exceed $100 per gross ton of such vessel or $14,000,000, whichever is lesser, except that where the United States can show that such discharge was the result of willful negligence or willful misconduct within the privity and knowledge of the owner, such owner or operator shall be liable to the United States Government for the full amount of such costs. Such costs shall constitute a maritime lien on such vessel which may be recovered in an action in rem in the district court of the United *Liability.*

173

States for any district within which any vessel may be found. The United States may also bring an action against the owner or operator of such vessel in any court of competent jurisdiction to recover such costs.

"(2) Except where an owner or operator of an onshore facility can prove that a discharge was caused solely by (A) an act of God, (B) an act of war, (C) negligence on the part of the United States Government, or (D) an act or omission of a third party without regard to whether any such act or omission was or was not negligent, or any combination of the foregoing clauses, such owner or operator of any such facility from which oil or a hazardous substance is discharged in violation of subsection (b)(2) of this section shall be liable to the United States Government for the actual costs incurred under subsection (c) for the removal of such oil or substance by the United States Government in an amount not to exceed $8,000,000, except that where the United States can show that such discharge was the result of willful negligence or willful misconduct within the privity and knowledge of the owner, such owner or operator shall be liable to the United States Government for the full amount of such costs. The United States may bring an action against the owner or operator of such facility in any court of competent jurisdiction to recover such costs. The Secretary is authorized, by regulation, after consultation with the Secretary of Commerce and the Small Business Administration, to establish reasonable and equitable classifications of those onshore facilities having a total fixed storage capacity of 1,000 barrels or less which he determines because of size, type, and location do not present a substantial risk of the discharge of oil or a hazardous substance in violation of subsection (b)(2) of this section, and apply with respect to such classifications differing limits of liability which may be less than the amount contained in this paragraph.

"(3) Except where an owner or operator of an offshore facility can prove that a discharge was caused solely by (A) an act of God, (B) an act of war, (C) negligence on the part of the United States Government, or (D) an act or omission of a third party without regard to whether any such act or omission was or was not negligent, or any combination of the foregoing clauses, such owner or operator of any such facility from which oil or a hazardous substance is discharged in violation of subsection (b)(2) of this section shall, notwithstanding any other provision of law, be liable to the United States Government for the actual costs incurred under subsection (c) for the removal of such oil or substance by the United States Government in an amount not to exceed $8,000,000, except that where the United States can show that such discharge was the result of willful negligence or willful misconduct within the privity and knowledge of the owner, such owner or operator shall be liable to the United States Government for the full amount of such costs. The United States may bring an action against the owner or operator of such a facility in any court of competent jurisdiction to recover such costs.

"(g) In any case where an owner or operator of a vessel, of an onshore facility, or of an offshore facility, from which oil or a hazardous substance is discharged in violation of subsection (b)(2) of this section, proves that such discharge of oil or hazardous substance was caused solely by an act or omission of a third party, or was caused solely by such an act or omission in combination with an act of God, an act of war, or negligence on the part of the United States Government, such third party shall, notwithstanding any other provision of law, be liable to the United States Government for the actual costs incurred under subsection (c) for removal of such oil or substance by the United States Government, except where such third party can

174

prove that such discharge was caused solely by (A) an act of God, (B) an act of war, (C) negligence on the part of the United States Government, or (D) an act or omission of another party without regard to whether such act or omission was or was not negligent, or any combination of the foregoing clauses. If such third party was the owner or operator of a vessel which caused the discharge of oil or a hazardous substance in violation of subsection (b)(2) of this section, the liability of such third party under this subsection shall not exceed $100 per gross ton of such vessel or $14,000,000, whichever is the lesser. In any other case the liability of such third party shall not exceed the limitation which would have been applicable to the owner or operator of the vessel or the onshore or offshore facility from which the discharge actually occurred if such owner or operator were liable. If the United States can show that the discharge of oil or a hazardous substance in violation of subsection (b)(2) of this section was the result of willful negligence or willful misconduct within the privity and knowledge of such third party, such third party shall be liable to the United States Government for the full amount of such removal costs. The United States may bring an action against the third party in any court of competent jurisdiction to recover such removal costs.

"(h) The liabilities established by this section shall in no way affect any rights which (1) the owner or operator of a vessel or of an onshore facility or an offshore facility may have against any third party whose acts may in any way have caused or contributed to such discharge, or (2) The United States Government may have against any third party whose actions may in any way have caused or contributed to the discharge of oil or hazardous substance.

"(i)(1) In any case where an owner or operator of a vessel or an onshore facility or an offshore facility from which oil or a hazardous substance is discharged in violation of subsection (b)(2) of this section acts to remove such oil or substance in accordance with regulations promulgated pursuant to this section, such owner or operator shall be entitled to recover the reasonable costs incurred in such removal upon establishing, in a suit which may be brought against the United States Government in the United States Court of Claims, that such discharge was caused solely by (A) an act of God, (B) an act of war, (C) negligence on the part of the United States Government, or (D) an act or omission of a third party without regard to whether such act or omission was or was not negligent, or of any combination of the foregoing causes. *(Removal costs, recovery.)*

"(2) The provisions of this subsection shall not apply in any case where liability is established pursuant to the Outer Continental Shelf Lands Act. *(67 Stat. 462. 43 USC 1331 note.)*

"(3) Any amount paid in accordance with a judgment of the United States Court of Claims pursuant to this section shall be paid from the funds established pursuant to subsection (k).

"(j)(1) Consistent with the National Contingency Plan required by subsection (c)(2) of this section, as soon as practicable after the effective date of this section, and from time to time thereafter, the President shall issue regulations consistent with maritime safety and with marine and navigation laws (A) establishing methods and procedures for removal of discharged oil and hazardous substances, (B) establishing criteria for the development and implementation of local and regional oil and hazardous substance removal contingency plans, (C) establishing procedures, methods, and equipment and other requirements for equipment to prevent discharges of oil and hazardous substances from vessels and from onshore facilities and offshore facilities, and to contain such discharges, and (D) governing the inspection of vessels carrying cargoes of oil and hazardous substances *(Regulations.)*

175

and the inspection of such cargoes in order to reduce the likelihood of discharges of oil from vessels in violation of this section.

"(2) Any owner or operator of a vessel or an onshore facility or an offshore facility and any other person subject to any regulation issued under paragraph (1) of this subsection who fails or refuses to comply with the provisions of any such regulations, shall be liable to a civil penalty of not more than $5,000 for each such violation. Each violation shall be a separate offense. The President may assess and compromise such penalty. No penalty shall be assessed until the owner, operator, or other person charged shall have been given notice and an opportunity for a hearing on such charge. In determining the amount of the penalty, or the amount agreed upon in compromise, the gravity of the violation, and the demonstrated good faith of the owner, operator, or other person charged in attempting to achieve rapid compliance, after notification of a violation, shall be considered by the President.

"(k) There is hereby authorized to be appropriated to a revolving fund to be established in the Treasury not to exceed $35,000,000 to carry out the provisions of subsections (c), (d), (i), and (l) of this section. Any other funds received by the United States under this section shall also be deposited in said fund for such purposes. All sums appropriated to, or deposited in, said fund shall remain available until expended.

"(l) The President is authorized to delegate the administration of this section to the heads of those Federal departments, agencies, and instrumentalities which he determines to be appropriate. Any moneys in the fund established by subsection (k) of this section shall be available to such Federal departments, agencies, and instrumentalities to carry out the provisions of subsections (c) and (i) of this section. Each such department, agency, and instrumentality, in order to avoid duplication of effort, shall, whenever appropriate, utilize the personnel, services, and facilities of other Federal departments, agencies, and instrumentalities.

"(m) Anyone authorized by the President to enforce the provisions of this section may, except as to public vessels, (A) board and inspect any vessel upon the navigable waters of the United States or the waters of the contiguous zone, (B) with or without a warrant arrest any person who violates the provisions of this section or any regulation issued thereunder in his presence or view, and (C) execute any warrant or other process issued by an officer or court of competent jurisdiction.

"(n) The several district courts of the United States are invested with jurisdiction for any actions, other than actions pursuant to subsection (i)(1), arising under this section. In the case of Guam and the Trust Territory of the Pacific Islands, such actions may be brought in the district court of Guam, and in the case of the Virgin Islands such actions may be brought in the district court of the Virgin Islands. In the case of American Samoa and the Trust Territory of the Pacific Islands, such actions may be brought in the District Court of the United States for the District of Hawaii and such court shall have jurisdiction of such actions. In the case of the Canal Zone, such actions may be brought in the United States District Court for the District of the Canal Zone.

"(o)(1) Nothing in this section shall affect or modify in any way the obligations of any owner or operator of any vessel, or of any owner or operator of any onshore facility or offshore facility to any person or agency under any provision of law for damages to any publicly owned or privately owned property resulting from a discharge of any oil or hazardous substance or from the removal of any such oil or hazardous substance.

"(2) Nothing in this section shall be construed as preempting any State or political subdivision thereof from imposing any requirement or liability with respect to the discharge of oil or hazardous substance into any waters within such State.

"(3) Nothing in this section shall be construed as affecting or modifying any other existing authority of any Federal department, agency, or instrumentality, relative to onshore or offshore facilities under this Act or any other provision of law, or to affect any State or local law not in conflict with this section.

"(p)(1) Any vessel over three hundred gross tons, including any barge of equivalent size, but not including any barge that is not self-propelled and that does not carry oil or hazardous substances as cargo or fuel, using any port or place in the United States or the navigable waters of the United States for any purpose shall establish and maintain under regulations to be prescribed from time to time by the President, evidence of financial responsibility of $100 per gross ton, or $14,000,000 whichever is the lesser, to meet the liability to the United States which such vessel could be subjected under this section. In cases where an owner or operator owns, operates, or charters more than one such vessel, financial responsibility need only be established to meet the maximum liability to which the largest of such vessels could be subjected. Financial responsibility may be established by any one of, or a combination of, the following methods acceptable to the President: (A) evidence of insurance, (B) surety bonds, (C) qualification as a self-insurer, or (D) other evidence of financial responsibility. Any bond filed shall be issued by a bonding company authorized to do business in the United States.

Certain vessels, financial responsibility.

"(2) The provisions of paragraph (1) of this subsection shall be effective April 3, 1971, with respect to oil and one year after the date of enactment of this section with respect to hazardous substances. The President shall delegate the responsibility to carry out the provisions of this subsection to the appropriate agency head within sixty days after the date of enactment of this section. Regulations necessary to implement this subsection shall be issued within six months after the date of enactment of this section.

Effective date.

"(3) Any claim for costs incurred by such vessel may be brought directly against the insurer or any other person providing evidence of financial responsibility as required under this subsection. In the case of any action pursuant to this subsection such insurer or other person shall be entitled to invoke all rights and defenses which would have been available to the owner or operator if an action had been brought against him by the claimant, and which would have been available to him if an action had been brought against him by the owner or operator.

"(4) Any owner or operator of a vessel subject to this subsection, who fails to comply with the provisions of this subsection or any regulation issued thereunder, shall be subject to a fine of not more than $10,000.

Penalty.

"(5) The Secretary of the Treasury may refuse the clearance required by section 4197 of the Revised Statutes of the United States, as amended (4 U.S.C. 91), to any vessel subject to this subsection, which does not have evidence furnished by the President that the financial responsibility provisions of paragraph (1) of this subsection have been complied with.

46 USC 91.

"(6) The Secretary of the Department in which the Coast Guard is operated may (A) deny entry to any port or place in the United States or the navigable waters of the United States, to, and (B) detain at the port or place in the United States from which it is about to depart for any other port or place in the United States, any vessel subject to this subsection, which upon request, does not produce evidence furnished by the President that the financial responsibility provisions of paragraph (1) of this subsection have been complied with.

"MARINE SANITATION DEVICES

Definitions.

"SEC. 312. (a) For the purpose of this section, the term—

"(1) 'new vessel' includes every description of watercraft or other artificial contrivance used, or capable of being used, as a means of transportation on the navigable waters, the construction of which is initiated after promulgation of standards and regulations under this section;

"(2) 'existing vessel' includes every description of watercraft or other artificial contrivance used, or capable of being used, as a means of transportation on the navigable waters, the construction of which is initiated before promulgation of standards and regulations under this section;

"(3) 'public vessel' means a vessel owned or bareboat chartered and operated by the United States, by a State or political subdivision thereof, or by a foreign nation, except when such vessel is engaged in commerce;

"(4) 'United States' includes the States, the District of Columbia, the Commonwealth of Puerto Rico, the Virgin Islands, Guam, American Samoa, the Canal Zone, and the Trust Territory of the Pacific Islands;

"(5) 'marine sanitation device' includes any equipment for installation on board a vessel which is designed to receive, retain, treat, or discharge sewage, and any process to treat such sewage;

"(6) 'sewage' means human body wastes and the wastes from toilets and other receptacles intended to receive or retain body wastes;

"(7) 'manufacturer' means any person engaged in the manufacturing, assembling, or importation of marine sanitation devices or of vessels subject to standards and regulations promulgated under this section;

"(8) 'person' means an individual, partnership, firm, corporation, or association, but does not include an individual on board a public vessel;

"(9) 'discharge' includes, but is not limited to, any spilling, leaking, pumping, pouring, emitting, emptying or dumping.

Federal standards of performance, promulgation.

"(b) (1) As soon as possible, after the enactment of this section and subject to the provisions of section 104(j) of this Act, the Administrator, after consultation with the Secretary of the department in which the Coast Guard is operating, after giving appropriate consideration to the economic costs involved, and within the limits of available technology, shall promulgate Federal standards of performance for marine sanitation devices (hereafter in this section referred to as 'standards') which shall be designed to prevent the discharge of untreated or inadequately treated sewage into or upon the navigable waters from new vessels and existing vessels, except vessels not equipped with installed toilet facilities. Such standards shall be consistent with maritime safety and the marine and navigation laws and regulations and shall be coordinated with the regulations issued under this subsection by the Secretary of the department in which the Coast Guard is operating. The Secretary of the department in which the Coast Guard is operating shall promulgate regulations, which are consistent with standards promulgated under this subsection and with maritime safety and the marine and navigation laws and regulations governing the design, construction, installation, and operation of any marine sanitation device on board such vessels.

178

"(2) Any existing vessel equipped with a marine sanitation device on the date of promulgation of initial standards and regulations under this section, which device is in compliance with such initial standards and regulations, shall be deemed in compliance with this section until such time as the device is replaced or is found not to be in compliance with such initial standards and regulations.

"(c)(1) Initial standards and regulations under this section shall become effective' for new vessels two years after promulgation; and for existing vessels five years after promulgation. Revisions of standards and regulations shall be effective upon promulgation, unless another effective date is specified, except that no revision shall take effect before the effective date of the standard or regulation being revised. Initial standards, effective dates.

"(2) The Secretary of the department in which the Coast Guard is operating with regard to his regulatory authority established by this section, after consultation with the Administrator, may distinguish among classes, type, and sizes of vessels as well as between new and existing vessels, and may waive applicability of standards and regulations as necessary or appropriate for such classes, types, and sizes of vessels (including existing vessels equipped with marine sanitation devices on the date of promulgation of the initial standards required by this section), and, upon application, for individual vessels. Waiver.

"(d) The provisions of this section and the standards and regulations promulgated hereunder apply to vessels owned and operated by the United States unless the Secretary of Defense finds that compliance would not be in the interest of national security. With respect to vessels owned and operated by the Department of Defense, regulations under the last sentence of subsection (b)(1) of this section and certifications under subsection (g)(2) of this section shall be promulgated and issued by the Secretary of Defense.

"(e) Before the standards and regulations under this section are promulgated, the Administrator and the Secretary of the department in which the Coast Guard is operating shall consult with the Secretary of State; the Secretary of Health, Education, and Welfare; the Secretary of Defense; the Secretary of the Treasury; the Secretary of Commerce; other interested Federal agencies; and the States and industries interested; and otherwise comply with the requirements of section 553 of title 5 of the United States Code. 80 Stat. 383.

"(f)(1) After the effective date of the initial standards and regulations promulgated under this section, no State or political subdivision thereof shall adopt or enforce any statute or regulation of such State or political subdivision with respect to the design, manufacture, or installation or use of any marine sanitation device on any vessel subject to the provisions of this section.

"(2) If, after promulgation of the initial standards and regulations and prior to their effective date, a vessel is equipped with a marine sanitation device in compliance with such standards and regulations and the installation and operation of such device is in accordance with such standards and regulations, such standards and regulations shall, for the purposes of paragraph (1) of this subsection, become effective with repect to such vessel on the date of such compliance.

"(3) After the effective date of the initial standards and regulations promulgated under this section, if any State determines that the protection and enhancement of the quality of some or all of the waters within such State require greater environmental protection, such State may completely prohibit the discharge from all vessels of any sewage, whether treated or not, into such waters, except that no such prohibition shall apply until the Administrator determines that adequate facilities for the safe and sanitary removal and treatment

179

of sewage from all vessels are reasonably available for such water to which such prohibition would apply. Upon application of the State, the Administrator shall make such determination within 90 days of the date of such application.

"(4) If the Administrator determines upon application by a State that the protection and enhancement of the quality of specified waters within such State requires such a prohibition, he shall by regulation completely prohibit the discharge from a vessel of any sewage (whether treated or not) into such waters.

Sales regulations.

"(g)(1) No manufacturer of a marine sanitation device shall sell, offer for sale, or introduce or deliver for introduction in interstate commerce, or import into the United States for sale or resale any marine sanitation device manufactured after the effective date of the standards and regulations promulgated under this section unless such device is in all material respects substantially the same as a test device certified under this subsection.

"(2) Upon application of the manufacturer, the Secretary of the department in which the Coast Guard is operating shall so certify a marine sanitation device if he determines, in accordance with the provisions of this paragraph, that it meets the appropriate standards and regulations promulgated under this section. The Secretary of the department in which the Coast Guard is operating shall test or require such testing of the device in accordance with procedures set forth by the Administrator as to standards of performance and for such other purposes as may be appropriate. If the Secretary of the department in which the Coast Guard is operating determines that the device is satisfactory from the standpoint of safety and any other requirements of maritime law or regulation, and after consideration of the design, installation, operation, material, or other appropriate factors, he shall certify the device. Any device manufactured by such manufacturer which is in all material respects substantially the same as the certified test device shall be deemed to be in conformity with the appropriate standards and regulations established under this section.

Recordkeeping; reports.

"(3) Every manufacturer shall establish and maintain such records, make such reports, and provide such information as the Administrator or the Secretary of the department in which the Coast Guard is operating may reasonably require to enable him to determine whether such manufacturer has acted or is acting in compliance with this section and regulations issued thereunder and shall, upon request of an officer or employee duly designated by the Administrator or the Secretary of the department in which the Coast Guard is operating, permit such officer or employee at reasonable times to have access to and copy such records. All information reported to or otherwise obtained by the Administrator or the Secretary of the department in which the Coast Guard is operating or their representatives pursuant to this subsection which contains or relates to a trade secret or other matter referred to in section 1905 of title 18 of the United States

62 Stat. 791.

Code shall be considered confidential for the purpose of that section, except that such information may be disclosed to other officers or employees concerned with carrying out this section. This paragraph shall not apply in the case of the construction of a vessel by an individual for his own use.

"(h) After the effective date of standards and regulations promulgated under this section, it shall be unlawful—

"(1) for the manufacturer of any vessel subject to such standards and regulations to manufacture for sale, to sell or offer for sale, or to distribute for sale or resale any such vessel unless it is equipped with a marine sanitation device which is in all material respects substantially the same as the appropriate test device certified pursuant to this section;

180

"(2) for any person, prior to the sale or delivery of a vessel subject to such standards and regulations to the ultimate purchaser, wrongfully to remove or render inoperative any certified marine sanitation device or element of design of such device installed in such vessel;

"(3) for any person to fail or refuse to permit access to or copying of records or to fail to make reports or provide information required under this section; and

"(4) for a vessel subject to such standards and regulations to operate on the navigable waters of the United States, if such vessel is not equipped with an operable marine sanitation device certified pursuant to this section.

"(i) The district courts of the United States shall have jurisdictions Jurisdiction. to restrain violations of subsection (g)(1) of this section and subsections (h)(1) through (3) of this section. Actions to restrain such violations shall be brought by, and in, the name of the United States. In case of contumacy or refusal to obey a subpena served upon any person under this subsection, the district court of the United States for any district in which such person is found or resides or transacts business, upon application by the United States and after notice to such person, shall have jurisdiction to issue an order requiring such person to appear and give testimony or to appear and produce documents, and any failure to obey such order of the court may be punished by such court as a contempt thereof.

"(j) Any person who violates subsection (g)(1) of this section or Penalties. clause (1) or (2) of subsection (h) of this section shall be liable to a civil penalty of not more than $5,000 for each violation. Any person who violates clause (4) of subsection (h) of this section or any regulation issued pursuant to this section shall be liable to a civil penalty of not more than $2,000 for each violation. Each violation shall be a separate offense. The Secretary of the department in which the Coast Guard is operating may assess and compromise any such penalty. No penalty shall be assessed until the person charged shall have been given notice and an opportunity for a hearing on such charge. In determining the amount of the penalty, or the amount agreed upon in compromise, the gravity of the violation, and the demonstrated good faith of the person charged in attempting to achieve rapid compliance, after notification of a violation, shall be considered by said Secretary.

"(k) The provisions of this section shall be enforced by the Secretary of the department in which the Coast Guard is operating and he may utilize by agreement, with or without reimbursement, law enforcement officers or other personnel and facilities of the Administrator, other Federal agencies, or the States to carry out the provisions of this section.

"(l) Anyone authorized by the Secretary of the department in which the Coast Guard is operating to enforce the provisions of this section may, except as to public vessels, (1) board and inspect any vessel upon the navigable waters of the United States and (2) execute any warrant or other process issued by an officer or court of competent jurisdiction.

"(m) In the case of Guam and the Trust Territory of the Pacific Islands, actions arising under this section may be brought in the district court of Guam, and in the case of the Virgin Islands such actions may be brought in the district court of the Virgin Islands. In the case of American Samoa and the Trust Territory of the Pacific Islands, such actions may be brought in the District Court of the United States for the District of Hawaii and such court shall have jurisdiction of such actions. In the case of the Canal Zone, such actions may be brought in the District Court for the District of the Canal Zone.

"SEC. 313. Each department, agency, or instrumentality of the executive, legislative, and judicial branches of the Federal Government (1) having jurisdiction over any property or facility, or (2) engaged in any activity resulting, or which may result, in the discharge or runoff of pollutants shall comply with Federal, State, interstate, and local requirements respecting control and abatement of pollution to the same extent that any person is subject to such requirements, including the payment of reasonable service charges. The President may exempt any effluent source of any department, agency, or instrumentality in the executive branch from compliance with any such a requirement if he determines it to be in the paramount interest of the United States to do so; except that no exemption may be granted from the requirements of section 306 or 307 of this Act. No such exemptions shall be granted due to lack of appropriation unless the President shall have specifically requested such appropriation as a part of the budgetary process and the Congress shall have failed to make available such requested appropriation. Any exemption shall be for a period not in excess of one year, but additional exemptions may be granted for periods of not to exceed one year upon the President's making a new determination. The President shall report each January to the Congress all exemptions from the requirements of this section granted during the preceding calendar year, together with his reason for granting such exemption.

Margin note: Exemption.

Margin note: Report to Congress.

"CLEAN LAKES

"SEC. 314. (a) Each State shall prepare or establish, and submit to the Administrator for his approval—

"(1) an identification and classification according to eutrophic condition of all publicly owned fresh water lakes in such State;

"(2) procedures, processes, and methods (including land use requirements), to control sources of pollution of such lakes; and

"(3) methods and procedures, in conjunction with appropriate Federal agencies, to restore the quality of such lakes.

"(b) The Administrator shall provide financial assistance to States in order to carry out methods and procedures approved by him under this section.

"(c)(1) The amount granted to any State for any fiscal year under this section shall not exceed 70 per centum of the funds expended by such State in such year for carrying out approved methods and procedures under this section.

"(2) There is authorized to be appropriated $50,000,000 for the fiscal year ending June 30, 1973; $100,000,000 for the fiscal year 1974; and $150,000,000 for the fiscal year 1975 for grants to States under this section which such sums shall remain available until expended. The Administrator shall provide for an equitable distribution of such sums to the States with approved methods and procedures under this section.

Margin note: Appropriations.

"NATIONAL STUDY COMMISSION

"SEC. 315. (a) There is established a National Study Commission, which shall make a full and complete investigation and study of all of the technological aspects of achieving, and all aspects of the total economic, social, and environmental effects of achieving or not achieving, the effluent limitations and goals set forth for 1983 in section 301(b)(2) of this Act.

"(b) Such Commission shall be composed of fifteen members, including five members of the Senate, who are members of the Public Works committee, appointed by the President of the Senate, five mem-

Margin note: Establishment.

bers of the House, who are members of the Public Works committee, appointed by the Speaker of the House, and five members of the public appointed by the President. The Chairman of such Commission shall be elected from among its members.

"(c) In the conduct of such study, the Commission is authorized to contract with the National Academy of Sciences and the National Academy of Engineering (acting through the National Research Council), the National Institute of Ecology, Brookings Institution, and other nongovernmental entities, for the investigation of matters within their competence. Contract authority.

"(d) The heads of the departments, agencies and instrumentalities of the executive branch of the Federal Government shall cooperate with the Commission in carrying out the requirements of this section, and shall furnish to the Commission such information as the Commission deems necessary to carry out this section.

"(e) A report shall be submitted to the Congress of the results of such investigation and study, together with recommendations, not later than three years after the date of enactment of this title. Report to Congress.

"(f) The members of the Commission who are not officers or employees of the United States, while attending conferences or meetings of the Commission or while otherwise serving at the request of the Chairman shall be entitled to receive compensation at a rate not in excess of the maximum rate of pay for grade GS–18, as provided in the General Schedule under section 5332 of title V of the United States Code, including traveltime and while away from their homes or regular places of business they may be allowed travel expenses, including per diem in lieu of subsistence as authorized by law (5 U.S.C. 73b–2) for persons in the Government service employed intermittently. 5 USC 5703 and note.

"(g) There is authorized to be appropriated, for use in carrying out this section, not to exceed $15,000,000. Appropriation.

"THERMAL DISCHARGES

"Sec. 316. (a) With respect to any point source otherwise subject to the provisions of section 301 or section 306 of this Act, whenever the owner or operator of any such source, after opportunity for public hearing, can demonstrate to the satisfaction of the Administrator (or, if appropriate, the State) that any effluent limitation proposed for the control of the thermal component of any discharge from such source will require effluent limitations more stringent than necessary to assure the projection and propagation of a balanced, indigenous population of shellfish, fish, and wildlife in and on the body of water into which the discharge is to be made, the Administrator (or, if appropriate, the State) may impose an effluent limitation under such sections for such plant, with respect to the thermal component of such discharge (taking into account the interaction of such thermal component with other pollutants), that will assure the protection and propagation of a balanced, indigenous population of shellfish, fish, and wildlife in and on that body of water.

"(b) Any standard established pursuant to section 301 or section 306 of this Act and applicable to a point source shall require that the location, design, construction, and capacity of cooling water intake structures reflect the best technology available for minimizing adverse environmental impact.

"(c) Notwithstanding any other provision of this Act, any point source of a discharge having a thermal component, the modification of which point source is commenced after the date of enactment of the Federal Water Pollution Control Act Amendments of 1972 and which, as modified, meets effluent limitations established under section 301 or, if more stringent, effluent limitations established under section 303 and which effluent limitations will assure protection and propagation of a balanced, indigenous population of shellfish, fish, and wildlife in or on the water into which the discharge is made, shall not be subject to any more stringent effluent limitation with respect to the thermal component of its discharge during a ten year period beginning on the date of completion of such modification or during the period of depreciation or amortization of such facility for the purpose of section 167 or 169 (or both) of the Internal Revenue Code of 1954, whichever period ends first.

68A Stat. 51;
85 Stat. 508.
83 Stat. 667.
26 USC 167,
169.

"FINANCING STUDY

84 Stat. 91.
33 USC 1151
note.
Report to
Congress.

"SEC. 317. (a) The Administrator shall continue to investigate and study the feasibility of alternate methods of financing the cost of preventing, controlling and abating pollution as directed in the Water Quality Improvement Act of 1970 (Public Law 91–224), including, but not limited to, the feasibility of establishing a pollution abatement trust fund. The results of such investigation and study shall be reported to the Congress not later than two years after enactment of this title, together with recommendations of the Administrator for financing the programs for preventing, controlling and abating pollution for the fiscal years beginning after fiscal year 1976, including any necessary legislation.

Appropriation.

"(b) There is authorized to be appropriated for use in carrying out this section, not to exceed $1,000,000.

"AQUACULTURE

"SEC. 318. (a) The Administrator is authorized, after public hearings, to permit the discharge of a specific pollutant or pollutants under controlled conditions associated with an approved aquaculture project under Federal or State supervision.

"(b) The Administrator shall by regulation, not later than January 1, 1974, establish any procedures and guidelines he deems necessary to carry out this section.

"TITLE IV—PERMITS AND LICENSES

"CERTIFICATION

"SEC. 401. (a)(1) Any applicant for a Federal license or permit to conduct any activity including, but not limited to, the construction or operation of facilities, which may result in any discharge into the navigable waters, shall provide the licensing or permitting agency a certification from the State in which the discharge originates or will originate, or, if appropriate, from the interstate water pollution control agency having jurisdiction over the navigable waters at the point where the discharge originates or will originate, that any such discharge will comply with the applicable provisions of sections 301, 302, 306, and 307 of this Act. In the case of any such activity for which there is not an applicable effluent limitation or other limitation under sections 301(b) and 302, and there is not an applicable standard under sections 306 and 307, the State shall so certify, except that any such certification shall not be deemed to satisfy section 511(c) of this Act. Such State or interstate agency shall establish procedures for public notice in the case of all applications for certification by it and, to the extent it deems appropriate, procedures for public hearings in connection with specific applications. In any case where a State or interstate agency has no authority to give such a certification, such certification shall be from the Administrator. If the State, interstate

184

agency, or Administrator, as the case may be, fails or refuses to act on a request for certification, within a reasonable period of time (which shall not exceed one year) after receipt of such request, the certification requirements of this subsection shall be waived with respect to such Federal application. No license or permit shall be granted until the certification required by this section has been obtained or has been waived as provided in the preceding sentence. No license or permit shall be granted if certification has been denied by the State, interstate agency, or the Administrator, as the case may be.

"(2) Upon receipt of such application and certification the licensing or permitting agency shall immediately notify the Administrator of such application and certification. Whenever such a discharge may affect, as determined by the Administrator, the quality of the waters of any other State, the Administrator within thirty days of the date of notice of application for such Federal license or permit shall so notify such other State, the licensing or permitting agency, and the applicant. If, within sixty days after receipt of such notification, such other State determines that such discharge will affect the quality of its waters so as to violate any water quality requirement in such State, and within such sixty-day period notifies the Administrator and the licensing or permitting agency in writing of its objection to the issuance of such license or permit and requests a public hearing on such objection, the licensing or permitting agency shall hold such a hearing. The Administrator shall at such hearing submit his evaluation and recommendations with respect to any such objection to the licensing or permitting agency. Such agency, based upon the recommendations of such State, the Administrator, and upon any additional evidence, if any, presented to the agency at the hearing, shall condition such license or permit in such manner as may be necessary to insure compliance with applicable water quality requirements. If the imposition of conditions cannot insure such compliance such agency shall not issue such license or permit.

"(3) The certification obtained pursuant to paragraph (1) of this subsection with respect to the construction of any facility shall fulfill the requirements of this subsection with respect to certification in connection with any other Federal license or permit required for the operation of such facility unless, after notice to the certifying State, agency, or Administrator, as the case may be, which shall be given by the Federal agency to whom application is made for such operating license or permit, the State, or if appropriate, the interstate agency or the Administrator, notifies such agency within sixty days after receipt of such notice that there is no longer reasonable assurance that there will be compliance with the applicable provisions of sections 301, 302, 306, and 307 of this Act because of changes since the construction license or permit certification was issued in (A) the construction or operation of the facility, (B) the characteristics of the waters into which such discharge is made, (C) the water quality criteria applicable to such waters or (D) applicable effluent limitations or other requirements. This paragraph shall be inapplicable in any case where the applicant for such operating license or permit has failed to provide the certifying State, or, if appropriate, the interstate agency or the Administrator, with notice of any proposed changes in the construction or operation of the facility with respect to which a construction license or permit has been granted, which changes may result in violation of section 301, 302, 306, or 307 of this Act.

"(4) Prior to the initial operation of any federally licensed or permitted facility or activity which may result in any discharge into the navigable waters and with respect to which a certification has been obtained pursuant to paragraph (1) of this subsection, which facility or activity is not subject to a Federal operating license or permit, the licensee or permittee shall provide an opportunity for such certifying State, or, if appropriate, the interstate agency or the Administrator to review the manner in which the facility or activity shall be operated or conducted for the purposes of assuring that applicable effluent limitations or other limitations or other applicable water quality requirements will not be violated. Upon notification by the certifying State, or if appropriate, the interstate agency or the Administrator that the operation of any such federally licensed or permitted facility or activity will violate applicable effluent limitations or other limitations or other water quality requirements such Federal agency may, after public hearing, suspend such license or permit. If such license or permit is suspended, it shall remain suspended until notification is received from the certifying State, agency, or Administrator, as the case may be, that there is reasonable assurance that such facility or activity will not violate the applicable provisions of section 301, 302, 306, or 307 of this Act.

"(5) Any Federal license or permit with respect to which a certification has been obtained under paragraph (1) of this subsection may be suspended or revoked by the Federal agency issuing such license or permit upon the entering of a judgment under this Act that such facility or activity has been operated in violation of the applicable provisions of section 301, 302, 306, or 307 of this Act.

"(6) No Federal agency shall be deemed to be an applicant for the purposes of this subsection.

"(7) Except with respect to a permit issued under section 402 of this Act, in any case where actual construction of a facility has been lawfully commenced prior to April 3, 1970, no certification shall be required under this subsection for a license or permit issued after April 3, 1970, to operate such facility, except that any such license or permit issued without certification shall terminate April 3, 1973, unless prior to such termination date the person having such license or permit submits to the Federal agency which issued such license or permit a certification and otherwise meets the requirements of this section.

"(b) Nothing in this section shall be construed to limit the authority of any department or agency pursuant to any other provision of law to require compliance with any applicable water quality requirements. The Administrator shall, upon the request of any Federal department or agency, or State or interstate agency, or applicant, provide, for the purpose of this section, any relevant information on applicable effluent limitations, or other limitations, standards, regulations, or requirements, or water quality criteria, and shall, when requested by any such department or agency or State or interstate agency, or applicant, comment on any methods to comply with such limitations, standards, regulations, requirements, or criteria.

"(c) In order to implement the provisions of this section, the Secretary of the Army, acting through the Chief of Engineers, is authorized, if he deems it to be in the public interest, to permit the use of spoil disposal areas under his jurisdiction by Federal licensees or permittees, and to make an appropriate charge for such use. Moneys received from such licensees or permittees shall be deposited in the Treasury as miscellaneous receipts.

"(d) Any certification provided under this section shall set forth any effluent limitations and other limitations, and monitoring require-

186

ments necessary to assure that any applicant for a Federal license or permit will comply with any applicable effluent limitations and other limitations, under section 301 or 302 of this Act, standard of performance under section 306 of this Act, or prohibition, effluent standard, or pretreatment standard under section 307 of this Act, and with any other appropriate requirement of State law set forth in such certification, and shall become a condition on any Federal license or permit subject to the provisions of this section.

"SEC. 402. (a) (1) Except as provided in sections 318 and 404 of this Act, the Administrator may, after opportunity for public hearing, issue a permit for the discharge of any pollutant, or combination of pollutants, notwithstanding section 301(a), upon condition that such discharge will meet either all applicable requirements under sections 301, 302, 306, 307, 308, and 403 of this Act, or prior to the taking of necessary implementing actions relating to all such requirements, such conditions as the Administrator determines are necessary to carry out the provisions of this Act. *(Permits, issuance.)*

"(2) The Administrator shall prescribe conditions for such permits to assure compliance with the requirements of paragraph (1) of this subsection, including conditions on data and information collection, reporting, and such other requirements as he deems appropriate.

"(3) The permit program of the Administrator under paragraph (1) of this subsection, and permits issued thereunder, shall be subject to the same terms, conditions, and requirements as apply to a State permit program and permits issued thereunder under subsection (b) of this section.

"(4) All permits for discharges into the navigable waters issued pursuant to section 13 of the Act of March 3, 1899, shall be deemed to be permits issued under this title, and permits issued under this title shall be deemed to be permits issued under section 13 of the Act of March 3, 1899, and shall continue in force and effect for their term unless revoked, modified, or suspended in accordance with the provisions of this Act. *(30 Stat. 1152. 33 USC 407.)*

"(5) No permit for a discharge into the navigable waters shall be issued under section 13 of the Act of March 3, 1899, after the date of enactment of this title. Each application for a permit under section 13 of the Act of March 3, 1899, pending on the date of enactment of this Act shall be deemed to be an application for a permit under this section. The Administrator shall authorize a State, which he determines has the capability of administering a permit program which will carry out the objective of this Act, to issue permits for discharges into the navigable waters within the jurisdiction of such State. The Administrator may exercise the authority granted him by the preceding sentence only during the period which begins on the date of enactment of this Act and ends either on the ninetieth day after the date of the first promulgation of guidelines required by section 304 (h) (2) of this Act, or the date of approval by the Administrator of a permit program for such State under subsection (b) of this section, whichever date first occurs, and no such authorization to a State shall extend beyond the last day of such period. Each such permit shall be subject to such conditions as the Administrator determines are necessary to carry out the provisions of this Act. No such permit shall issue if the Administrator objects to such issuance.

"(b) At any time after the promulgation of the guidelines required by subsection (h) (2) of section 304 of this Act, the Governor of each State desiring to administer its own permit program for discharges *(State permit programs.)*

into navigable waters within its jurisdiction may submit to the Administrator a full and complete description of the program it proposes to establish and administer under State law or under an interstate compact. In addition, such State shall submit a statement from the attorney general (or the attorney for those State water pollution control agencies which have independent legal counsel), or from the chief legal officer in the case of an interstate agency, that the laws of such State, or the interstate compact, as the case may be, provide Approval conditions. adequate authority to carry out the described program. The Administrator shall approve each such submitted program unless he determines that adequate authority does not exist:

"(1) To issue permits which—

"(A) apply, and insure compliance with, any applicable requirements of sections 301, 302, 306, 307, and 403;

"(B) are for fixed terms not exceeding five years; and

"(C) can be terminated or modified for cause including, but not limited to, the following:

"(i) violation of any condition of the permit;

"(ii) obtaining a permit by misrepresentation, or failure to disclose fully all relevant facts;

"(iii) change in any condition that requires either a temporary or permanent reduction or elimination of the permitted discharge;

"(D) control the disposal of pollutants into wells;

"(2)(A) To issue permits which apply, and insure compliance with, all applicable requirements of section 308 of this Act, or

"(B) To inspect, monitor, enter, and require reports to at least the same extent as required in section 308 of this Act;

"(3) To insure that the public, and any other State the waters of which may be affected, receive notice of each application for a permit and to provide an opportunity for public hearing before a ruling on each such application;

"(4) To insure that the Administrator receives notice of each application (including a copy thereof) for a permit;

"(5) To insure that any State (other than the permitting State), whose waters may be affected by the issuance of a permit may submit written recommendations to the permitting State (and the Administrator) with respect to any permit application and, if any part of such written recommendations are not accepted by the permitting State, that the permitting State will notify such affected State (and the Administrator) in writing of its failure to so accept such recommendations together with its reasons for so doing;

"(6) To insure that no permit will be issued if, in the judgment of the Secretary of the Army acting through the Chief of Engineers, after consultation with the Secretary of the department in which the Coast Guard is operating, anchorage and navigation of any of the navigable waters would be substantially impaired thereby;

"(7) To abate violations of the permit or the permit program, including civil and criminal penalties and other ways and means of enforcement;

"(8) To insure that any permit for a discharge from a publicly owned treatment works includes conditions to require adequate notice to the permitting agency of (A) new introductions into such works of pollutants from any source which would be a new source as defined in section 306 if such source were discharging pollutants, (B) new introductions of pollutants into such works from a source which would be subject to section 301 if it were discharging such pollutants, or (C) a substantial change in volume or character of pollutants being introduced into such works by a source introducing pollutants into such works at the time of issuance of the permit. Such notice shall include information on the quality and quantity of effluent to be introduced into such treatment works and any anticipated impact of such change in the quantity or quality of effluent to be discharged from such publicly owned treatment works; and

"(9) To insure that any industrial user of any publicly owned treatment works will comply with sections 204(b), 307, and 308.

"(c) (1) Not later than ninety days after the date on which a State has submitted a program (or revision thereof) pursuant to subsection (b) of this section, the Administrator shall suspend the issuance of permits under subsection (a) of this section as to those navigable waters subject to such program unless he determines that the State permit program does not meet the requirements of subsection (b) of this section or does not conform to the guidelines issued under section 304(h)(2) of this Act. If the Administrator so determines, he shall notify the State of any revisions or modifications necessary to conform to such requirements or guidelines.

"(2) Any State permit program under this section shall at all times be in accordance with this section and guidelines promulgated pursuant to section 304(h)(2) of this Act.

"(3) Whenever the Administrator determines after public hearing that a State is not administering a program approved under this section in accordance with requirements of this section, he shall so notify the State and, if appropriate corrective action is not taken within a reasonable time, not to exceed ninety days, the Administrator shall withdraw approval of such program. The Administrator shall not withdraw approval of any such program unless he shall first have notified the State, and made public, in writing, the reasons for such withdrawal. State permit program, approval withdrawal.

"(d) (1) Each State shall transmit to the Administrator a copy of each permit application received by such State and provide notice to the Administrator of every action related to the consideration of such permit application, including each permit proposed to be issued by such State. Administrator, notification.

"(2) No permit shall issue (A) if the Administrator within ninety days of the date of his notification under subsection (b)(5) of this section objects in writing to the issuance of such permit, or (B) if the Administrator within ninety days of the date of transmittal of the proposed permit by the State objects in writing to the issuance of such permit as being outside the guidelines and requirements of this Act.

"(3) The Administrator may, as to any permit application, waive paragraph (2) of this subsection. Waiver authority.

"(e) In accordance with guidelines promulgated pursuant to subsection (h)(2) of section 304 of this Act, the Administrator is authorized to waive the requirements of subsection (d) of this section at the time he approves a program pursuant to subsection (b) of this section for any category (including any class, type, or size within such category) of point sources within the State submitting such program.

"(f) The Administrator shall promulgate regulations establishing categories of point sources which he determines shall not be subject to the requirements of subsection (d) of this section in any State with a program approved pursuant to subsection (b) of this section. The Administrator may distinguish among classes, types, and sizes within any category of point sources. Point sources, categories.

"(g) Any permit issued under this section for the discharge of pollutants into the navigable waters from a vessel or other floating craft shall be subject to any applicable regulations promulgated by the Secretary of the department in which the Coast Guard is operating,

189

establishing specifications for safe transportation, handling, carriage, storage, and stowage of pollutants.

"(h) In the event any condition of a permit for discharges from a treatment works (as defined in section 212 of this Act) which is publicly owned is violated, a State with a program approved under subsection (b) of this section or the Administrator, where no State program is approved, may proceed in a court of competent jurisdiction to restrict or prohibit the introduction of any pollutant into such treatment works by a source not utilizing such treatment works prior to the finding that such condition was violated.

"(i) Nothing in this section shall be construed to limit the authority of the Administrator to take action pursuant to section 309 of this Act.

Public information.

"(j) A copy of each permit application and each permit issued under this section shall be available to the public. Such permit application or permit, or portion thereof, shall further be available on request for the purpose of reproduction.

"(k) Compliance with a permit issued pursuant to this section shall be deemed compliance, for purposes of sections 309 and 505, with sections 301, 302, 306, 307, and 403, except any standard imposed under section 307 for a toxic pollutant injurious to human health. Until December 31, 1974, in any case where a permit for discharge has been applied for pursuant to this section, but final administrative disposition of such application has not been made, such discharge shall not be a violation of (1) section 301, 306, or 402 of this Act, or (2) section 13

30 Stat. 1152.
33 USC 407.

of the Act of March 3, 1899, unless the Administrator or other plaintiff proves that final administrative disposition of such application has not been made because of the failure of the applicant to furnish information reasonably required or requested in order to process the application. For the 180-day period beginning on the date of enactment of the

Ante, p. 816.

Federal Water Pollution Control Act Amendments of 1972, in the case of any point source discharging any pollutant or combination of pollutants immediately prior to such date of enactment which source is not subject to section 13 of the Act of March 3, 1899, the discharge by such source shall not be a violation of this Act if such a source applies for a permit for discharge pursuant to this section within such 180-day period.

"OCEAN DISCHARGE CRITERIA

"SEC. 403. (a) No permit under section 402 of this Act for a discharge into the territorial sea, the waters of the contiguous zone, or the oceans shall be issued, after promulgation of guidelines established under subsection (c) of this section, except in compliance with such guidelines. Prior to the promulgation of such guidelines, a permit may be issued under such section 402 if the Administrator determines it to be in the public interest.

"(b) The requirements of subsection (d) of section 402 of this Act may not be waived in the case of permits for discharges into the territorial sea.

Guidelines.

"(c)(1) The Administrator shall, within one hundred and eighty days after enactment of this Act (and from time to time thereafter), promulgate guidelines for determining the degradation of the waters of the territorial seas, the contiguous zone, and the oceans, which shall include:

"(A) the effect of disposal of pollutants on human health or welfare, including but not limited to plankton, fish, shellfish, wildlife, shorelines, and beaches;

"(B) the effect of disposal of pollutants on marine life including the transfer, concentration, and dispersal of pollutants or their byproducts through biological, physical, and chemical processes; changes in marine ecosystem diversity, productivity, and stability; and species and community population changes;

"(C) the effect of disposal, of pollutants on esthetic, recreation, and economic values;

"(D) the persistence and permanence of the effects of disposal of pollutants;

"(E) the effect of the disposal at varying rates, of particular volumes and concentrations of pollutants;

"(F) other possible locations and methods of disposal or recycling of pollutants including land-based alternatives; and

"(G) the effect on alternate uses of the oceans, such as mineral exploitation and scientific study.

"(2) In any event where insufficient information exists on any proposed discharge to make a reasonable judgment on any of the guidelines established pursuant to this subsection no permit shall be issued under section 402 of this Act.

Prohibition.

"PERMITS FOR DREDGED OR FILL MATERIAL

"SEC. 404. (a) The Secretary of the Army, acting through the Chief of Engineers, may issue permits, after notice and opportunity for public hearings for the discharge of dredged or fill material into the navigable waters at specified disposal sites.

Notice, hearing opportunity.

"(b) Subject to subsection (c) of this section, each such disposal site shall be specified for each such permit by the Secretary of the Army (1) through the application of guidelines developed by the Administrator, in conjunction with the Secretary of the Army, which guidelines shall be based upon criteria comparable to the criteria applicable to the territorial seas, the contiguous zone, and the ocean under section 403(c), and (2) in any case where such guidelines under clause (1) alone would prohibit the specification of a site, through the application additionally of the economic impact of the site on navigation and anchorage.

"(c) The Administrator is authorized to prohibit the specification (including the withdrawal of specification) of any defined area as a disposal site, and he is authorized to deny or restrict the use of any defined area for specification (including the withdrawal of specification) as a disposal site, whenever he determines, after notice and opportunity for public hearings, that the discharge of such materials into such area will have an unacceptable adverse effect on municipal water supplies, shellfish beds and fishery areas (including spawning and breeding areas), wildlife, or recreational areas. Before making such determination, the Administrator shall consult with the Secretary of the Army. The Administrator shall set forth in writing and make public his findings and his reasons for making any determination under this subsection.

Disposal site, specification prohibition.

Findings of Administrator, publication.

"DISPOSAL OF SEWAGE SLUDGE

"SEC. 405. (a) Notwithstanding any other provision of this Act or of any other law, in any case where the disposal of sewage sludge resulting from the operation of a treatment works as defined in section 212 of this Act (including the removal of in-place sewage sludge from one location and its deposit at another location) would result in any pollutant from such sewage sludge entering the navigable waters, such disposal is prohibited except in accordance with a permit issued by the Administrator under this section.

"(b) The Administrator shall issue regulations governing the issuance of permits for the disposal of sewage sludge subject to this section. Such regulations shall require the application to such disposal of each criterion, factor, procedure, and requirement applicable to a permit issued under section 402 of this title, as the Administrator determines necessary to carry out the objective of this Act.

State permit program. "(c) Each State desiring to administer its own permit program for disposal of sewage sludge within its jurisdiction may do so if upon submission of such program the Administrator determines such program is adequate to carry out the objective of this Act.

"TITLE V—GENERAL PROVISIONS

"ADMINISTRATION

Regulations. "SEC. 501. (a) The Administrator is authorized to prescribe such regulations as are necessary to carry out his functions under this Act.

"(b) The Administrator, with the consent of the head of any other agency of the United States, may utilize such officers and employees of such agency as may be found necessary to assist in carrying out the purposes of this Act.

Recordkeeping. "(c) Each recipient of financial assistance under this Act shall keep such records as the Administrator shall prescribe, including records which fully disclose the amount and disposition by such recipient of the proceeds of such assistance, the total cost of the project or undertaking in connection with which such assistance is given or used, and the amount of that portion of the cost of the project or undertaking supplied by other sources, and such other records as will facilitate an effective audit.

Audit. "(d) The Administrator and the Comptroller General of the United States, or any of their duly authorized representatives, shall have access, for the purpose of audit and examination, to any books, documents, papers, and records of the recipients that are pertinent to the grants received under this Act.

Awards. "(e)(1) It is the purpose of this subsection to authorize a program which will provide official recognition by the United States Government to those industrial organizations and political subdivisions of States which during the preceding year demonstrated an outstanding technological achievement or an innovative process, method, or device in their waste treatment and pollution abatement programs. The

Application, regulations. Administrator shall, in consultation with the appropriate State water pollution control agencies, establish regulations under which such recognition may be applied for and granted, except that no applicant shall be eligible for an award under this subsection if such applicant is not in total compliance with all applicable water quality requirements under this Act, or otherwise does not have a satisfactory record with respect to environmental quality.

"(2) The Administrator shall award a certificate or plaque of suitable design to each industrial organization or political subdivision which qualifies for such recognition under regulations established under this subsection.

Publication in Federal Register. "(3) The President of the United States, the Governor of the appropriate State, the Speaker of the House of Representatives, and the President pro tempore of the Senate shall be notified of the award by the Administrator and the awarding of such recognition shall be published in the Federal Register.

EPA personnel, detail to State agencies. "(f) Upon the request of a State water pollution control agency, personnel of the Environmental Protection Agency may be detailed to such agency for the purpose of carrying out the provisions of this Act.

192

"SEC. 502. Except as otherwise specifically provided, when used in this Act:

"(1) The term 'State water pollution control agency' means the State agency designated by the Governor having responsibility for enforcing State laws relating to the abatement of pollution.

"(2) The term 'interstate agency' means an agency of two or more States established by or pursuant to an agreement or compact approved by the Congress, or any other agency of two or more States, having substantial powers or duties pertaining to the control of pollution as determined and approved by the Administrator.

"(3) The term 'State' means a State, the District of Columbia, the Commonwealth of Puerto Rico, the Virgin Islands, Guam, American Samoa, and the Trust Territory of the Pacific Islands.

"(4) The term 'municipality' means a city, town, borough, county, parish, district, association, or other public body created by or pursuant to State law and having jurisdiction over disposal of sewage, industrial wastes, or other wastes, or an Indian tribe or an authorized Indian tribal organization, or a designated and approved management agency under section 208 of this Act.

"(5) The term 'person' means an individual, corporation, partnership, association, State, municipality, commission, or political subdivision of a State, or any interstate body.

"(6) The term 'pollutant' means dredged spoil, solid waste, incinerator residue, sewage, garbage, sewage sludge, munitions, chemical wastes, biological materials, radioactive materials, heat, wrecked or discarded equipment, rock, sand, cellar dirt and industrial, municipal, and agricultural waste discharged into water. This term does not mean (A) 'sewage from vessels' within the meaning of section 312 of this Act; or (B) water, gas, or other material which is injected into a well to facilitate production of oil or gas, or water derived in association with oil or gas production and disposed of in a well, if the well used either to facilitate production or for disposal purposes is approved by authority of the State in which the well is located, and if such State determines that such injection or disposal will not result in the degradation of ground or surface water resources.

"(7) The term 'navigable waters' means the waters of the United States, including the territorial seas.

"(8) The term 'territorial seas' means the belt of the seas measured from the line of ordinary low water along that portion of the coast which is in direct contact with the open sea and the line marking the seaward limit of inland waters, and extending seaward a distance of three miles.

"(9) The term 'contiguous zone' means the entire zone established or to be established by the United States under article 24 of the Convention of the Territorial Sea and the Contiguous Zone.   15 UST 1606.

"(10) The term 'ocean' means any portion of the high seas beyond the contiguous zone.

"(11) The term 'effluent limitation' means any restriction established by a State or the Administrator on quantities, rates, and concentrations of chemical, physical, biological, and other constituents which are discharged from point sources into navigable waters, the waters of the contiguous zone, or the ocean, including schedules of compliance.

"(12) The term 'discharge of a pollutant' and the term 'discharge of pollutants' each means (A) any addition of any pollutant to navigable waters from any point source, (B) any addition of any pollutant to the waters of the contiguous zone or the ocean from any point source other than a vessel or other floating craft.

"(13) The term 'toxic pollutant' means those pollutants, or combinations of pollutants, including disease-causing agents, which after discharge and upon exposure, ingestion, inhalation or assimilation into any organism, either directly from the environment or indirectly by ingestion through food chains, will, on the basis of information available to the Administrator, cause death, disease, behavioral abnormalities, cancer, genetic mutations, physiological malfunctions (including malfunctions in reproduction) or physical deformations, in such organisms or their offspring.

"(14) The term 'point source' means any discernible, confined and discrete conveyance, including but not limited to any pipe, ditch, channel, tunnel, conduit, well, discrete fissure, container, rolling stock, concentrated animal feeding operation, or vessel or other floating craft, from which pollutants are or may be discharged.

"(15) The term 'biological monitoring' shall mean the determination of the effects on aquatic life, including accumulation of pollutants in tissue, in receiving waters due to the discharge of pollutants (A) by techniques and procedures, including sampling of organisms representative of appropriate levels of the food chain appropriate to the volume and the physical, chemical, and biological characteristics of the effluent, and (B) at appropriate frequencies and locations.

"(16) The term 'discharge' when used without qualification includes a discharge of a pollutant, and a discharge of pollutants.

"(17) The term 'schedule of compliance' means a schedule of remedial measures including an enforceable sequence of actions or operations leading to compliance with an effluent limitation, other limitation, prohibition, or standard.

"(18) The term 'industrial user' means those industries identified in the Standard Industrial Classification Manual, Bureau of the Budget, 1967, as amended and supplemented, under the category 'Division D—Manufacturing' and such other classes of significant waste producers as, by regulation, the Administrator deems appropriate.

"(19) The term 'pollution' means the man-made or man-induced alteration of the chemical, physical, biological, and radiological integrity of water.

"WATER POLLUTION CONTROL ADVISORY BOARD

Establishment.

"SEC. 503. (a) (1) There is hereby established in the Environmental Protection Agency a Water Pollution Control Advisory Board, composed of the Administrator or his designee, who shall be Chairman, and nine members appointed by the President, none of whom shall be Federal officers or employees. The appointed members, having due regard for the purposes of this Act, shall be selected from among representatives of various State, interstate, and local governmental agencies, of public or private interests contributing to, affected by, or concerned with pollution, and of other public and private agencies, organizations, or groups demonstrating an active interest in the field of pollution prevention and control, as well as other individuals who are expert in this field.

Term.

"(2) (A) Each member appointed by the President shall hold office for a term of three years, except that (i) any member appointed to fill a vacancy occurring prior to the expiration of the term for which his predecessor was appointed shall be appointed for the remainder of such term, and (ii) the terms of office of the members first taking office after June 30, 1956, shall expire as follows: three at the end of one year after such date, three at the end of two years after such date, and three at the end of three years after such date, as designated by the President at the time of appointment, and (iii) the term of any member under

194

the preceding provisions shall be extended until the date on which his successor's appointment is effective. None of the members appointed by the President shall be eligible for reappointment within one year after the end of his preceding term.

"(B) The members of the Board who are not officers or employees of the United States. while attending conferences or meetings of the Board or while otherwise serving at the request of the Administrator, shall be entitled to receive compensation at a rate to be fixed by the Administrator. but not exceeding $100 per diem, including travel-time, and while away from their homes or regular places of business they may be allowed travel expenses, including per diem in lieu of subsistence. as authorized by law (5 U.S.C. 73b–2) for persons in the Government service employed intermittently. Compensation. 80 Stat. 499; 83 Stat. 190. 5 USC 5703, 5707 and notes.

"(b) The Board shall advise, consult with, and make recommendations to the Administrator on matters of policy relating to the activities and functions of the Administrator under this Act.

"(c) Such clerical and technical assistance as may be necessary to discharge the duties of the Board shall be provided from the personnel of the Environmental Protection Agency.

"EMERGENCY POWERS

"SEC. 504. Notwithstanding any other provision of this Act, the Administrator upon receipt of evidence that a pollution source or combination of sources is presenting an imminent and substantial endangerment to the health of persons or to the welfare of persons where such endangerment is to the livelihood of such persons, such as inability to market shellfish, may bring suit on behalf of the United States in the appropriate district court to immediately restrain any person causing or contributing to the alleged pollution to stop the discharge of pollutants causing or contributing to such pollution or to take such other action as may be necessary.

"CITIZEN SUITS

"SEC. 505. (a) Except as provided in subsection (b) of this section, any citizen may commence a civil action on his own behalf—

"(1) against any person (including (i) the United States, and (ii) any other governmental instrumentality or agency to the extent permitted by the eleventh amendment to the Constitution) who is alleged to be in violation of (A) an effluent standard or limitation under this Act or (B) an order issued by the Administrator or a State with respect to such a standard or limitation, or USC prec. title 1.

"(2) against the Administrator where there is alleged a failure of the Administrator to perform any act or duty under this Act which is not discretionary with the Administrator.

The district courts shall have jurisdiction, without regard to the amount in controversy or the citizenship of the parties, to enforce such an effluent standard or limitation, or such an order, or to order the Administrator to perform such act or duty, as the case may be, and to apply any appropriate civil penalties under section 309(d) of this Act. District court jurisdiction.

"(b) No action may be commenced— Notice and filing.

"(1) under subsection (a)(1) of this section—

"(A) prior to sixty days after the plaintiff has given notice of the alleged violation (i) to the Administrator, (ii) to the State in which the alleged violation occurs, and (iii) to any alleged violator of the standard, limitation, or order, or

"(B) if the Administrator or State has commenced and is diligently prosecuting a civil or criminal action in a court of the United States, or a State to require compliance with the standard, limitation, or order, but in any such action in a court of the United States any citizen may intervene as a matter of right.

"(2) under subsection (a)(2) of this section prior to sixty days after the plaintiff has given notice of such action to the Administrator,

except that such action may be brought immediately after such notification in the case of an action under this section respecting a violation of sections 306 and 307(a) of this Act. Notice under this subsection shall be given in such manner as the Administrator shall prescribe by regulation.

Notice, regulation.

"(c)(1) Any action respecting a violation by a discharge source of an effluent standard or limitation or an order respecting such standard or limitation may be brought under this section only in the judicial district in which such source is located.

"(2) In such action under this section, the Administrator, if not a party, may intervene as a matter of right.

Litigation costs.

"(d) The court, in issuing any final order in any action brought pursuant to this section, may award costs of litigation (including reasonable attorney and expert witness fees) to any party, whenever the court determines such award is appropriate. The court may, if a temporary restraining order or preliminary injunction is sought, require the filing of a bond or equivalent security in accordance with the Federal Rules of Civil Procedure.

28 USC app.

"(e) Nothing in this section shall restrict any right which any person (or class of persons) may have under any statute or common law to seek enforcement of any effluent standard or limitation or to seek any other relief (including relief against the Administrator or a State agency).

"Effluent standard or limitation under this Act."

"(f) For purposes of this section, the term 'effluent standard or limitation under this Act' means (1) effective July 1, 1973, an unlawful act under subsection (a) of section 301 of this Act; (2) an effluent limitation or other limitation under section 301 or 302 of this Act; (3) standard of performance under section 306 of this Act; (4) prohibition, effluent standard or pretreatment standards under section 307 of this Act; (5) certification under section 401 of this Act; or (6) a permit or condition thereof issued under section 402 of this Act, which is in effect under this Act (including a requirement applicable by reason of section 313 of this Act).

"Citizen."

"(g) For the purposes of this section the term 'citizen' means a person or persons having an interest which is or may be adversely affected.

State Governor, civil action.

"(h) A Governor of a State may commence a civil action under subsection (a), without regard to the limitations of subsection (b) of this section, against the Administrator where there is alleged a failure of the Administrator to enforce an effluent standard or limitation under this Act the violation of which is occurring in another State and is causing an adverse effect on the public health or welfare in his State, or is causing a violation of any water quality requirement in his State.

"APPEARANCE

"SEC. 506. The Administrator shall request the Attorney General to appear and represent the United States in any civil or criminal action instituted under this Act to which the Administrator is a party. Unless the Attorney General notifies the Administrator within a reasonable time, that he will appear in a civil action, attorneys who are officers or employees of the Environmental Protection Agency shall appear and represent the United States in such action.

"Sec. 507. (a) No person shall fire, or in any other way discriminate against. or cause to be fired or discriminated against, any employee or any authorized representative of employees by reason of the fact that such employee or representative has filed, instituted, or caused to be filed or instituted any proceeding under this Act, or has testified or is about to testify in any proceeding resulting from the administration or enforcement of the provisions of this Act.

"(b) Any employee or a representative of employees who believes that he has been fired or otherwise discriminated against by any person in violation of subsection (a) of this section may, within thirty days after such alleged violation occurs, apply to the Secretary of Labor for a review of such firing or alleged discrimination. A copy of the application shall be sent to such person who shall be the respondent. Upon receipt of such application, the Secretary of Labor shall cause such investigation to be made as he deems appropriate. Such investigation shall provide an opportunity for a public hearing at the request of any party to such review to enable the parties to present information relating to such alleged violation. The parties shall be given written notice of the time and place of the hearing at least five days prior to the hearing. Any such hearing shall be of record and shall be subject to section 554 of title 5 of the United States Code. Upon receiving the report of such investigation, the Secretary of Labor shall make findings of fact. If he finds that such violation did occur, he shall issue a decision, incorporating an order therein and his findings, requiring the party committing such violation to take such affirmative action to abate the violation as the Secretary of Labor deems appropriate, including, but not limited to, the rehiring or reinstatement of the employee or representative of employees to his former position with compensation. If he finds that there was no such violation, he shall issue an order denying the application. Such order issued by the Secretary of Labor under this subparagraph shall be subject to judicial review in the same manner as orders and decisions of the Administrator are subject to judicial review under this Act.

"(c) Whenever an order is issued under this section to abate such violation, at the request of the applicant, a sum equal to the aggregate amount of all costs and expenses (including the attorney's fees), as determined by the Secretary of Labor, to have been reasonably incurred by the applicant for, or in connection with, the institution and prosecution of such proceedings, shall be assessed against the person committing such violation.

"(d) This section shall have no application to any employee who, acting without direction from his employer (or his agent) deliberately violates any prohibition of effluent limitation or other limitaton under section 301 or 302 of this Act, standards of performance under section 306 of this Act, effluent standard, prohibition or pretreatment standard under section 307 of this Act, or any other prohibition or limitation established under this Act.

"(e) The Administrator shall conduct continuing evaluations of potential loss or shifts of employment which may result from the issuance of any effluent limitation or order under this Act, including, where appropriate, investigating threatened plant closures or reductions in employment allegedly resulting from such limitation or order. Any employee who is discharged or laid-off, threatened with discharge or lay-off, or otherwise discriminated against by any person because of the alleged results of any effluent limitation or order issued under this Act, or any representative of such employee, may request the Administrator to conduct a full investigation of the matter. The Administra-

Review; hearing opportunity.

81 Stat. 54.

Judicial review.

Employment reduction, investigation.

197

tor shall thereupon investigate the matter and, at the request of any party, shall hold public hearings on not less than five days notice, and shall at such hearings require the parties, including the employer involved, to present information relating to the actual or potential effect of such limitation or order on employment and on any alleged discharge, lay-off, or other discrimination and the detailed reasons or justification therefor. Any such hearing shall be of record and shall

81 Stat. 54.

be subject to section 554 of title 5 of the United States Code. Upon receiving the report of such investigation, the Administrator shall make findings of fact as to the effect of such effluent limitation or order on employment and on the alleged discharge, lay-off, or discrimination and shall make such recommendations as he deems appro-

Information, availability.

priate. Such report, findings, and recommendations shall be available to the public. Nothing in this subsection shall be construed to require or authorize the Administrator to modify or withdraw any effluent limitation or order issued under this Act.

"FEDERAL PROCUREMENT

Prohibition.

"SEC. 508. (a) No Federal agency may enter into any contract with any person, who has been convicted of any offense under section 309 (c) of this Act, for the procurement of goods, materials, and services if such contract is to be performed at any facility at which the violation which gave rise to such conviction occurred, and if such facility is owned, leased, or supervised by such person. The prohibition in the preceding sentence shall continue until the Administrator certifies that the condition giving rise to such conviction has been corrected.

"(b) The Administrator shall establish procedures to provide all Federal agencies with the notification necessary for the purposes of subsection (a) of this section.

Implementation, Presidential order.

"(c) In order to implement the purposes and policy of this Act to protect and enhance the quality of the Nation's water, the President shall, not more than one hundred and eighty days after enactment of this Act, cause to be issued an order (1) requiring each Federal agency authorized to enter into contracts and each Federal agency which is empowered to extend Federal assistance by way of grant, loan, or contract to effectuate the purpose and policy of this Act in such contracting or assistance activities, and (2) setting forth procedures, sanctions, penalties, and such other provisions, as the President determines necessary to carry out such requirement.

Exemption, notification to Congress.

"(d) The President may exempt any contract, loan, or grant from all or part of the provisions of this section where he determines such exemption is necessary in the paramount interest of the United States and he shall notify the Congress of such exemption.

Annual report to Congress.

"(e) The President shall annually report to the Congress on measures taken in compliance with the purpose and intent of this section, including, but not limited to, the progress and problems associated with such compliance.

"ADMINISTRATIVE PROCEDURE AND JUDICIAL REVIEW

Administrator, subpena power.

"SEC. 509. (a) (1) For purposes of obtaining information under section 305 of this Act, or carrying out section 507(e) of this Act, the Administrator may issue subpenas for the attendance and testimony of witnesses and the production of relevant papers, books, and docu-

Trade secrets, protection.

ments, and he may administer oaths. Except for effluent data, upon a showing satisfactory to the Administrator that such papers, books, documents, or information or particular part thereof, if made public, would divulge trade secrets or secret processes, the Administrator

198

shall consider such record, report, or information or particular portion thereof confidential in accordance with the purposes of section 1905 of title 18 of the United States Code, except that such paper, book, document, or information may be disclosed to other officers, employees, or authorized representatives of the United States concerned with carrying out this Act, or when relevant in any proceeding under this Act. Witnesses summoned shall be paid the same fees and mileage that are paid witnesses in the courts of the United States. In case of contumacy or refusal to obey a subpena served upon any person under this subsection, the district court of the United States for any district in which such person is found or resides or transacts business, upon application by the United States and after notice to such person, shall have jurisdiction to issue an order requiring such person to appear and give testimony before the Administrator, to appear and produce papers, books, and documents before the Administrator, or both, and any failure to obey such order of the court may be punished by such court as a contempt thereof. 62 Stat. 791.

Witnesses.

"(2) The district courts of the United States are authorized, upon application by the Administrator, to issue subpenas for attendance and testimony of witnesses and the production of relevant papers, books, and documents, for purposes of obtaining information under sections 304 (b) and (c) of this Act. Any papers, books, documents, or other information or part thereof, obtained by reason of such a subpena shall be subject to the same requirements as are provided in paragraph (1) of this subsection. U.S. district courts, subpenas..

"(b)(1) Review of the Administrator's action (A) in promulgating any standard of performance under section 306, (B) in making any determination pursuant to section 306(b)(1)(C), (C) in promulgating any effluent standard, prohibition, or treatment standard under section 307, (D) in making any determination as to a State permit program submitted under section 402(b), (E) in approving or promulgating any effluent limitation or other limitation under section 301, 302, or 306, and (F) in issuing or denying any permit under section 402, may be had by any interested person in the Circuit Court of Appeals of the United States for the Federal judicial district in which such person resides or transacts such business upon application by such person. Any such application shall be made within ninety days from the date of such determination, approval, promulgation, issuance or denial, or after such date only if such application is based solely on grounds which arose after such ninetieth day. Judicial review.

"(2) Action of the Administrator with respect to which review could have been obtained under paragraph (1) of this subsection shall not be subject to judicial review in any civil or criminal proceeding for enforcement.

"(c) In any judicial proceeding brought under subsection (b) of this section in which review is sought of a determination under this Act required to be made on the record after notice and opportunity for hearing, if any party applies to the court for leave to adduce additional evidence, and shows to the satisfaction of the court that such additional evidence is material and that there were reasonable grounds for the failure to adduce such evidence in the proceeding before the Administrator, the court may order such additional evidence (and evidence in rebuttal thereof) to be taken before the Administrator, in such manner and upon such terms and conditions as the court may deem proper. The Administrator may modify his findings as to the facts, or make new findings, by reason of the additional evidence so taken and he shall file such modified or new findings, and his recommendation, if any, for the modification or setting aside of his original determination, with the return of such additional evidence. Additional evidence; new findings.

"STATE AUTHORITY

"Sec. 510. Except as expressly provided in this Act, nothing in this Act shall (1) preclude or deny the right of any State or political subdivision thereof or interstate agency to adopt or enforce (A) any standard or limitation respecting discharges of pollutants, or (B) any requirement respecting control or abatement of pollution; except that if an effluent limitation, or other limitation, effluent standard, prohibition, pretreatment standard, or standard of performance is in effect under this Act, such State or political subdivision or interstate agency may not adopt or enforce any effluent limitation, or other limitation, effluent standard, prohibition, pretreatment standard, or standard of performance which is less stringent than the effluent limitation, or other limitation, effluent standard, prohibition, pretreatment standard, or standard of performance under this Act; or (2) be construed as impairing or in any manner affecting any right or jurisdiction of the States with respect to the waters (including boundary waters) of such States.

"OTHER AFFECTED AUTHORITY

"Sec. 511. (a) This Act shall not be construed as (1) limiting the authority or functions of any officer or agency of the United States under any other law or regulation not inconsistent with this Act; (2) affecting or impairing the authority of the Secretary of the Army (A) to maintain navigation or (B) under the Act of March 3, 1899 (30 Stat. 1112); except that any permit issued under section 404 of this Act shall be conclusive as to the effect on water quality of any discharge resulting from any activity subject to section 10 of the Act of March 3, 1899, or (3) affecting or impairing the provisions of any treaty of the United States.

30 Stat. 1151.
33 USC 403.

"(b) Discharges of pollutants into the navigable waters subject to the Rivers and Harbors Act of 1910 (36 Stat. 593; 33 U.S.C. 421) and the Supervisory Harbors Act of 1888 (25 Stat. 209; 33 U.S.C. 441–451b) shall be regulated pursuant to this Act, and not subject to such Act of 1910 and the Act of 1888 except as to effect on navigation and anchorage.

72 Stat. 970.

"(c)(1) Except for the provision of Federal financial assistance for the purpose of assisting the construction of publicly owned treatment works as authorized by section 201 of this Act, and the issuance of a permit under section 402 of this Act for the discharge of any pollutant by a new source as defined in section 306 of this Act, no action of the Administrator taken pursuant to this Act shall be deemed a major Federal action significantly affecting the quality of the human environment within the meaning of the National Environmental Policy Act of 1969 (83 Stat. 852); and

42 USC 4321
note.

"(2) Nothing in the National Environmental Policy Act of 1969 (83 Stat. 852) shall be deemed to—

"(A) authorize any Federal agency authorized to license or permit the conduct of any activity which may result in the discharge of a pollutant into the navigable waters to review any effluent limitation or other requirement established pursuant to this Act or the adequacy of any certification under section 401 of this Act; or

"(B) authorize any such agency to impose, as a condition precedent to the issuance of any license or permit, any effluent limitation other than any such limitation established pursuant to this Act.

"SEC. 512. If any provision of this Act, or the application of any provision of this Act to any person or circumstance, is held invalid, the application of such provision to other persons or circumstances, and the remainder of this Act, shall not be affected thereby.

"LABOR STANDARDS

"SEC. 513. The Administrator shall take such action as may be necessary to insure that all laborers and mechanics employed by contractors or subcontractors on treatment works for which grants are made under this Act shall be paid wages at rates not less than those prevailing for the same type of work on similar construction in the immediate locality, as determined by the Secretary of Labor, in accordance with the Act of March 3, 1931, as amended, known as the Davis-Bacon Act (46 Stat. 1494; 40 U.S.C., sec. 276a through 276a–5). 49 Stat. 1011; The Secretary of Labor shall have, with respect to the labor standards 78 Stat. 238. specified in this subsection, the authority and functions set forth in Reorganization Plan Numbered 14 of 1950 (15 F.R. 3176) and section 64 Stat. 1267. 2 of the Act of June 13, 1934, as amended (48 Stat. 948; 40 U.S.C. 63 Stat. 108. 276c).

"PUBLIC HEALTH AGENCY COORDINATION

"SEC. 514. The permitting agency under section 402 shall assist the applicant for a permit under such section in coordinating the requirements of this Act with those of the appropriate public health agencies.

"EFFLUENT STANDARDS AND WATER QUALITY INFORMATION ADVISORY COMMITTEE

"SEC. 515. (a) (1) There is established on Effluent Standards and Establishment. Water Quality Information Advisory Committee, which shall be composed of a Chairman and eight members who shall be appointed by the Administrator within sixty days after the date of enactment of this Act.

"(2) All members of the Committee shall be selected from the scien- Membership. tific community, qualified by education, training, and experience to provide, assess, and evaluate scientific and technical information on effluent standards and limitations.

"(3) Members of the Committee shall serve for a term of four years, Term. and may be reappointed.

"(b) (1) No later than one hundred and eighty days prior to the Proposed date on which the Administrator is required to publish any proposed regulations, regulations required by section 304(b) of this Act, any proposed public hearing. standard of performance for new sources required by section 306 of this Act, or any proposed toxic effluent standard required by section 307 of this Act, he shall transmit to the Committee a notice of intent to propose such regulations. The Chairman of the Committee within ten days after receipt of such notice may publish a notice of a public hearing by the Committee, to be held within thirty days.

"(2) No later than one hundred and twenty days after receipt of such notice, the Committee shall transmit to the Administrator such scientific and technical information as is in its possession, including that presented at any public hearing, related to the subject matter contained in such notice.

"(3) Information so transmitted to the Administrator shall constitute a part of the administrative record and comments on any proposed regulations or standards as information to be considered with other comments and information in making any final determinations.

"(4) In preparing information for transmittal, the Committee shall avail itself of the technical and scientific services of any Federal agency, including the United States Geological Survey and any national environmental laboratories which may be established.

"(c)(1) The Committee shall appoint and prescribe the duties of a Secretary, and such legal counsel as it deems necessary. The Committee shall appoint such other employees as it deems necessary to exercise and fulfill its powers and responsibilities. The compensation of all employees appointed by the Committee shall be fixed in accordance with chapter 51 and subchapter III of chapter 53 of title V of the United States Code.

"(2) Members of the Committee shall be entitled to receive compensation at a rate to be fixed by the President but not in excess of the maximum rate of pay for grade GS–18, as provided in the General Schedule under section 5332 of title V of the United States Code.

"(d) Five members of the Committee shall constitute a quorum, and official actions of the Committee shall be taken only on the affirmative vote of at least five members. A special panel composed of one or more members upon order of the Committee shall conduct any hearing authorized by this section and submit the transcript of such hearing to the entire Committee for its action thereon.

"(e) The Committee is authorized to make such rules as are necessary for the orderly transaction of its business.

"REPORTS TO CONGRESS

"SEC. 516. (a) Within ninety days following the convening of each session of Congress, the Administrator shall submit to the Congress a report, in addition to any other report required by this Act, on measures taken toward implementing the objective of this Act, including, but not limited to, (1) the progress and problems associated with developing comprehensive plans under section 102 of this Act, area-wide plans under section 208 of this Act, basin plans under section 209 of this Act, and plans under section 303(e) of this Act; (2) a summary of actions taken and results achieved in the field of water pollution control research, experiments, studies, and related matters by the Administrator and other Federal agencies and by other persons and agencies under Federal grants or contracts; (3) the progress and problems associated with the development of effluent limitations and recommended control techniques; (4) the status of State programs, including a detailed summary of the progress obtained as compared to that planned under State program plans for development and enforcement of water quality requirements; (5) the identification and status of enforcement actions pending or completed under such Act during the preceding year; (6) the status of State, interstate, and local pollution control programs established pursuant to, and assisted by, this Act; (7) a summary of the results of the survey required to be taken under section 210 of this Act; (8) his activities including recommendations under sections 109 through 111 of this Act; and (9) all reports and recommendations made by the Water Pollution Control Advisory Board.

"(b) The Administrator, in cooperation with the States, including water pollution control agencies and other water pollution control planning agencies, shall make (1) a detailed estimate of the cost of carrying out the provisions of this Act; (2) a detailed estimate, biennially revised, of the cost of construction of all needed publicly owned treatment works in all of the States and of the cost of construction of all needed publicly owned treatment works in each of the States; (3) a comprehensive study of the economic impact on affected units of government of the cost of installation of treatment facilities; and (4) a comprehensive analysis of the national requirements for and the cost of treating municipal, industrial, and other effluent to attain the water quality objectives as established by this Act or applicable State law. The Administrator shall submit such detailed estimate and such comprehensive study of such cost to the Congress no later than February 10 of each odd-numbered year. Whenever the Administrator, pursuant to this subsection, requests and receives an estimate of cost from a State, he shall furnish copies of such estimate together with such detailed estimate to Congress.

"GENERAL AUTHORIZATION

"SEC. 517. There are authorized to be appropriated to carry out this Act, other than sections 104, 105, 106(a), 107, 108, 112, 113, 114, 115, 206, 207, 208 (f) and (h), 209, 304, 311 (c), (d), (i), (l), and (k), 314, 315, and 317, $250,000,000 for the fiscal year ending June 30, 1973, $300,000,000 for the fiscal year ending June 30, 1974, and $350,000,000 for the fiscal year ending June 30, 1975.

"SHORT TITLE

"SEC. 518. This Act may be cited as the 'Federal Water Pollution Control Act'."

AUTHORIZATIONS FOR FISCAL YEAR 1972

SEC. 3. (a) There is authorized to be appropriated for the fiscal year ending June 30, 1972, and to exceed $11,000,000 for the purpose of carrying out section 5(n) (other than for salaries and related expenses) of the Federal Water Pollution Control Act as it existed immediately prior to the date of the enactment of the Federal Water Pollution Control Act Amendments of 1972.

84 Stat. 113.
33 USC 1155.

(b) There is hereby authorized to be appropriated for the fiscal year ending June 30, 1972, and to exceed $350,000,000 for the purpose of making grants under section 8 of the Federal Water Pollution Control Act as it existed immediately prior to the date of the enactment of the Federal Water Pollution Control Act Amendments of 1972.

79 Stat. 903;
80 Stat. 1248.
33 USC 1158.

(c) The Federal share of all grants made under section 8 of the Federal Water Pollution Control Act as it existed immediately prior to the date of enactment of the Federal Water Pollution Control Act Amendments of 1972 from sums herein and heretofore authorized for the fiscal year ending June 30, 1972, shall be that authorized by section 202 of such Act as established by the Federal Water Pollution Control Act Amendments of 1972.

Ante, p. 834.

(d) Sums authorized by this section shall be in addition to any amounts heretofore authorized for such fiscal year for sections 5(n) and 8 of the Federal Water Pollution Control Act as it existed immediately prior to the date of enactment of the Federal Water Pollution Control Act Amendments of 1972.

SAVINGS PROVISION

SEC. 4. (a) No suit, action, or other proceeding lawfully commenced by or against the Administrator or any other officer or employee of the United States in his official capacity or in relation to the discharge of his official duties under the Federal Water Pollution Control Act as in effect immediately prior to the date of enactment of this Act shall abate by reason of the taking effect of the amendment made by section 2 of this Act. The court may, on its own motion or that of any party

Ante, p. 816.

made at any time within twelve months after such taking effect, allow the same to be maintained by or against the Administrator or such officer or employee.

(b) All rules, regulations, orders, determinations, contracts, certifications, authorizations, delegations, or other actions duly issued, made, or taken by or pursuant to the Federal Water Pollution Control Act as in effect immediately prior to the date of enactment of this Act, and pertaining to any functions, powers, requirements, and duties under the Federal Water Pollution Control Act as in effect immediately prior to the date of enactment of this Act, shall continue in full force and effect after the date of enactment of this Act until modified or rescinded in accordance with the Federal Water Pollution Control Act as amended by this Act.

(c) The Federal Water Pollution Control Act as in effect immediately prior to the date of enactment of this Act shall remain applicable to all grants made from funds authorized for the fiscal year ending June 30, 1972, and prior fiscal years, including any increases in the monetary amount of any such grant which may be paid from authorizations for fiscal years beginning after June 30, 1972, except as specifically otherwise provided in section 202 of the Federal Water Pollution Control Act as amended by this Act and in subsection (c) of section 3 of this Act.

70 Stat. 498;
84 Stat. 91.
33 USC 1151
note.

Ante, p. 834.

OVERSIGHT STUDY

Report to
Congress.

SEC. 5. In order to assist the Congress in the conduct of oversight responsibilities the Comptroller General of the United States shall conduct a study and review of the research, pilot, and demonstration programs related to prevention and control of water pollution, including waste treatment and disposal techniques, which are conducted, supported, or assisted by any agency of the Federal Government pursuant to any Federal law or regulation and assess conflicts between, and the coordination and efficacy of, such programs, and make a report to the Congress thereon by October 1, 1973.

INTERNATIONAL TRADE STUDY

SEC. 6. (a) The Secretary of Commerce, in cooperation with other interested Federal agencies and with representatives of industry and the public, shall undertake immediately an investigation and study to determine—

(1) the extent to which pollution abatement and control programs will be imposed on, or voluntarily undertaken by, United States manufacturers in the near future and the probable short- and long-range effects of the costs of such programs (computed to the greatest extent practicable on an industry-by-industry basis) on (A) the production costs of such domestic manufacturers, and (B) the market prices of the goods produced by them;

(2) the probable extent to which pollution abatement and control programs will be implemented in foreign industrial nations in the near future and the extent to which the production costs (computed to the greatest extent practicable on an industry-by-industry basis) of foreign manufacturers will be affected by the costs of such programs;

(3) the probable competitive advantage which any article manufactured in a foreign nation will likely have in relation to a comparable article made in the United States if that foreign nation—

(A) does not require its manufacturers to implement pollution abatement and control programs.

(B) requires a lesser degree of pollution abatement and control in its programs, or

(C) in any way reimburses or otherwise subsidizes its manufacturers for the costs of such program;

(4) alternative means by which any competitive advantage accruing to the products of any foreign nation as a result of any factor described in paragraph (3) may be (A) accurately and quickly determined, and (B) equalized, for example, by the imposition of a surcharge or duty, on a foreign product in an amount necessary to compensate for such advantage; and

(5) the impact, if any, which the imposition of a compensating tariff of other equalizing measure may have in encouraging foreign nations to implement pollution and abatement control programs.

(b) The Secretary shall make an initial report to the President and Congress within six months after the date of enactment of this section of the results of the study and investigation carried out pursuant to this section and shall make additional reports thereafter at such times as he deems appropriate taking into account the development of relevant data, but not less than once every twelve months.

Report to President and Congress.

INTERNATIONAL AGREEMENTS

SEC. 7. The President shall undertake to enter into international agreements to apply uniform standards of performance for the control of the discharge and emission of pollutants from new sources, uniform controls over the discharge and emission of toxic pollutants, and uniform controls over the discharge of pollutants into the ocean. For this purpose the President shall negotiate multilateral treaties, conventions, resolutions, or other agreements, and formulate, present, or support proposals at the United Nations and other appropriate international forums.

LOANS TO SMALL BUSINESS CONCERNS FOR WATER POLLUTION CONTROL FACILITY

SEC. 8. (a) Section 7 of the Small Business Act is amended by inserting at the end thereof a new subsection as follows:

"(g)(1) The Administration also is empowered to make loans (either directly or in cooperation with banks or other lenders through agreements to participate on an immediate or deferred basis) to assist any small business concern in affecting additions to or alterations in the equipment, facilities (including the construction of pretreatment facilities and interceptor sewers), or methods of operation of such concern to meet water pollution control requirements established under the Federal Water Pollution Control Act, if the Administration determines that such concern is likely to suffer substantial economic injury without assistance under this subsection.

"(2) Any such loan—

"(A) shall be made in accordance with provisions applicable to loans made pursuant to subsection (b)(5) of this section, except as otherwise provided in this subsection;

"(B) shall be made only if the applicant furnishes the Administration with a statement in writing from the Environmental Protection Agency or, if appropriate, the State, that such addi-

72 Stat. 387; 81 Stat. 268. 15 USC 636.

Ante, p. 816.

84 Stat. 1633.

205

tions or alterations are necessary and adequate to comply with requirements established under the Federal Water Pollution Control Act.

Ante, p. 816.
Regulations, promulgation.

"(3) The Administrator of the Environmental Protection Agency shall, as soon as practicable after the date of enactment of the Federal Water Pollution Control Act Amendments of 1972 and not later than one hundred and eighty days thereafter, promulgate regulations establishing uniform rules for the issuance of statements for the purpose of paragraph (2) (B) of this subsection.

Ante, p. 382.
80 Stat. 132.
15 USC 633.

"(4) There is authorized to be appropriated to the disaster loan fund established pursuant to section 4(c) of this Act not to exceed $800,000,000 solely for the purpose of carrying out this subsection."

(b) Section 4(c) (1) (A) of the Small Business Act is amended by striking out "and 7(c) (2)" and inserting in lieu thereof "7(c) (2), and 7(g)"

ENVIRONMENTAL COURT

Report to Congress.

SEC. 9. The President, acting through the Attorney General, shall make a full and complete investigation and study of the feasibility of establishing a separate court, or court system, having jurisdiction over environmental matters and shall report the results of such investigation and study together with his recommendations to Congress not later than one year after the date of enactment of this Act.

NATIONAL POLICIES AND GOALS STUDY

Report to Congress.

SEC. 10. The President shall make a full and complete investigation and study of all of the national policies and goals established by law for the purpose of determining what the relationship should be between these policies and goals, taking into account the resources of the Nation. He shall report the results of such investigation and study together with his recommendations to Congress not later than two years after the date of enactment of this Act. There is authorized

Appropriation.

to be appropriated not to exceed $5,000,000 to carry out the purposes of this section.

EFFICIENCY STUDY

Report to Congress.

SEC. 11. The President shall conduct a full and complete investigation and study of ways and means of utilizing in the most effective manner all of the various resources, facilities, and personnel of the Federal Government in order most efficiently to carry out the objective of the Federal Water Pollution Control Act. He shall utilize in conducting such investigation and study, the General Accounting Office. He shall report the results of such investigation and study together with his recommendations to Congress not later than two hundred and seventy days after the date of enactment of this Act.

ENVIRONMENTAL FINANCING

Citation of title.
Environmental Financing Authority, establishment.

SEC. 12. (a) This section may be cited as the "Environmental Financing Act of 1972".

(b) There is hereby created a body corporate to be known as the Environmental Financing Authority, which shall have succession until dissolved by Act of Congress. The Authority shall be subject to the general supervision and direction of the Secretary of the Treasury. The Authority shall be an instrumentality of the United States Government and shall maintain such offices as may be necessary or appropriate in the conduct of its business.

206

(c) The purpose of this section is to assure that inability to borrow necessary funds on reasonable terms does not prevent any State or local public body from carrying out any project for construction of waste treatment works determined eligible for assistance pursuant to subsection (e) of this section.

(d)(1) The Authority shall have a Board of Directors consisting of five persons, one of whom shall be the Secretary of the Treasury or his designee as Chairman of the Board, and four of whom shall be appointed by the President from among the officers or employees of the Authority or of any department or agency of the United States Government.

Board of Directors.

(2) The Board of Directors shall meet at the call of its Chairman. The Board shall determine the general policies which shall govern the operations of the Authority. The Chairman of the Board shall select and effect the appointment of qualified persons to fill the offices as may be provided for in the bylaws, with such executive functions, powers, and duties as may be prescribed by the bylaws or by the Board of Directors, and such persons shall be the executive officers of the Authority and shall discharge all such executive functions, powers, and duties. The members of the Board, as such, shall not receive compensation for their services.

(e)(1) Until July 1, 1975, the Authority is authorized to make commitments to purchase, and to purchase on terms and conditions determined by the Authority, any obligation or participation therein which is issued by a State or local public body to finance the non-Federal share of the cost of any project for the construction of waste treatment works which the Administrator of the Environmental Protection Agency has determined to be eligible for Federal financial assistance under the Federal Water Pollution Control Act.

State or local obligations, purchase.

Ante, p. 816.

(2) No commitment shall be entered into, and no purchase shall be made, unless the Administrator of the Environmental Protection Agency (A) has certified that the public body is unable to obtain on reasonable terms sufficient credit to finance its actual needs; (B) has approved the project as eligible under the Federal Water Pollution Control Act; and (C) has agreed to guarantee timely payment of principal and interest on the obligation. The Administrator is authorized to guarantee such timely payments and to issue regulations as he deems necessary and proper to protect such guarantees. Appropriations are hereby authorized to be made to the Administrator in such sums as are necessary to make payments under such guarantees, and such payments are authorized to be made from such appropriations.

Appropriation.

(3) No purchase shall be made of obligations issued to finance projects, the permanent financing of which occurred prior to the enactment of this section.

Restriction.

(4) Any purchase by the Authority shall be upon such terms and conditions as to yield a return at a rate determined by the Secretary of the Treasury taking into consideration (A) the current average yield on outstanding marketable obligations of the United States of comparable maturity or in its stead whenever the Authority has sufficient of its own long-term obligations outstanding, the current average yield on outstanding obligations of the Authority of comparable maturity; and (B) the market yields on municipal bonds.

(5) The Authority is authorized to charge fees for its commitments and other services adequate to cover all expenses and to provide for the accumulation of reasonable contingency reserves and such fees shall be included in the aggregate project costs.

Fees.

(f) To provide initial capital to the Authority the Secretary of the Treasury is authorized to advance the funds necessary for this purpose. Each such advance shall be upon such terms and conditions

Advances, interest.

as to yield a return at a rate not less than a rate determined by the Secretary of the Treasury taking into consideration the current average yield on outstanding marketable obligations of the United States of comparable maturities. Interest payments on such advances may be deferred, at the discretion of the Secretary, but any such deferred payments shall themselves bear interest at the rate specified in this section. There is authorized to be appropriated not to exceed $100,000,000, which shall be available for the purposes of this subsection.

(g) (1) The Authority is authorized, with the approval of the Secretary of the Treasury, to issue and have outstanding obligations having such maturities and bearing such rate or rates of interest as may be determined by the Authority. Such obligations may be redeemable at the option of the Authority before maturity in such manner as may be stipulated therein.

(2) As authorized in appropriation Acts, and such authorizations may be without fiscal year limitation, the Secretary of the Treasury may in his discretion purchase or agree to purchase any obligations issued pursuant to paragraph (1) of this subsection, and for such purpose the Secretary of the Treasury is authorized to use as a public debt transaction the proceeds of the sale of any securities hereafter issued under the Second Liberty Bond Act, as now or hereafter in force, and the purposes for which securities may be issued under the Second Liberty Bond Act as now or hereafter in force, are extended to include such purchases. Each purchase of obligations by the Secretary of the Treasury under this subsection shall be upon such terms and conditions as to yield a return at a rate not less than a rate determined by the Secretary of the Treasury, taking into consideration the current average yield on outstanding marketable obligations of the United States of comparable maturities. The Secretary of the Treasury may sell, upon such terms and conditions and at such price or prices as he shall determine, any of the obligations acquired by him under this paragraph. All purchases and sales by the Secretary of the Treasury of such obligations under this paragraph shall be treated as public debt transactions of the United States.

(h) The Secretary of the Treasury is authorized and directed to make annual payments to the Authority in such amounts as are necessary to equal the amount by which the dollar amount of interest expense accrued by the Authority on account of its obligations exceeds the dollar amount of interest income accrued by the Authority on account of obligations purchased by it pursuant to subsection (e) of this section.

(i) The Authority shall have power—

(1) to sue and be sued, complain and defend, in its corporate name;

(2) to adopt, alter, and use a corporate seal, which shall be judicially noticed;

(3) to adopt, amend, and repeal bylaws, rules, and regulations as may be necessary for the conduct of its business;

(4) to conduct its business, carry on its operations, and have offices and exercise the powers granted by this section in any State without regard to any qualification or similar statute in any State;

(5) to lease, purchase, or otherwise acquire, own, hold, improve, use, or otherwise deal in and with any property, real, personal, or mixed, or any interest therein, wherever situated;

(6) to accept gifts or donations of services, or of property, real, personal, or mixed, tangible or intangible, in aid of any of the purposes of the Authority;

(7) to sell, convey, mortgage, pledge, lease, exchange, and otherwise dispose of its property and assets;

(8) to appoint such officers, attorneys, employees, and agents as may be required, to define their duties, to fix and to pay such compensation for their services as may be determined, subject to the civil service and classification laws, to require bonds for them and pay the premium thereof; and

(9) to enter into contracts, to execute instruments, to incur liabilities, and to do all things as are necessary or incidental to the proper management of its affairs and the proper conduct of its business.

(j) The Authority, its property, its franchise, capital, reserves, surplus, security holdings, and other funds, and its income shall be exempt from all taxation now or hereafter imposed by the United States or by any State or local taxing authority; except that (A) any real property and any tangible personal property of the Authority shall be subject to Federal, State, and local taxation to the same extent according to its value as other such property is taxed, and (B) any and all obligations issued by the Authority shall be subject both as to principal and interest to Federal, State, and local taxation to the same extent as the obligations of private corporations are taxed. Tax exemption; exceptions.

(k) All obligations issued by the Authority shall be lawful investments, and may be accepted as security for all fiduciary, trust, and public funds, the investment or deposit of which shall be under authority or control of the United States or of any officer or officers thereof. All obligations issued by the Authority pursuant to this section shall be deemed to be exempt securities within the meaning of laws administered by the Securities and Exchange Commission, to the same extent as securities which are issued by the United States.

(l) In order to furnish obligations for delivery by the Authority, the Secretary of the Treasury is authorized to prepare such obligations in such form as the Authority may approve, such obligations when prepared to be held in the Treasury subject to delivery upon order by the Authority. The engraved plates, dies, bed pieces, and so forth, executed in connection therewith, shall remain in the custody of the Secretary of the Treasury. The Authority shall reimburse the Secretary of the Treasury for any expenditures made in the preparation, custody, and delivery of such obligations.

(m) The Authority shall, as soon as practicable after the end of each fiscal year, transmit to the President and the Congress an annual report of its operations and activities. Annual report to President and Congress.

(n) The sixth sentence of the seventh paragraph of section 5136 of the Revised Statutes, as amended (12 U.S.C. 24), is amended by inserting "or obligations of the Environmental Financing Authority" immediately after "or obligations, participations, or other instruments of or issued by the Federal National Mortgage Association or the Government National Mortgage Association".

(o) The budget and audit provisions of the Government Corporation Control Act (31 U.S.C. 846) shall be applicable to the Environmental Financing Authority in the same manner as they are applied to the wholly owned Government corporations. 59 Stat. 597.

(p) Section 3689 of the Revised Statutes, as amended (31 U.S.C. 711), is further amended by adding a new paragraph following the last paragraph appropriating moneys for the purposes under the Treasury Department to read as follows:

"Payment to the Environmental Financing Authority: For payment to the Environmental Financing Authority under subsection (h) of the Environmental Financing Act of 1972." Ante, p. 901.

SEC. 13. No person in the United States shall on the ground of sex be excluded from participation in, be denied the benefits of, or be subjected to discrimination under any program or activity receiving Federal assistance under this Act, the Federal Water Pollution Control Act, or the Environmental Financing Act. This section shall be enforced through agency provisions and rules similar to those already established, with respect to racial and other discrimination, under title VI of the Civil Rights Act of 1964. However, this remedy is not exclusive and will not prejudice or cut off any other legal remedies available to a discriminatee.

Ante, pp. 816, 899.

78 Stat. 252.
42 USC 2000d.

CARL ALBERT

Speaker of the House of Representatives.

FRANK E. MOSS

acting President of the Senate pro tempore.

IN THE SENATE OF THE UNITED STATES,
October 18 (legislative day, October 17), 1972.

The Senate having proceeded to reconsider the bill (S. 2770) entitled "An Act to amend the Federal Water Pollution Control Act," returned by the President of the United States with his objections, to the Senate, in which it originated, it was
Resolved, That the said bill pass, two-thirds of the Senators present having voted in the affirmative.
Attest:

FRANCIS R. VALEO

Secretary.

By: Darrell St. Claire
Assistant Secretary.

I certify that this Act originated in the Senate.

FRANCIS R. VALEO

Secretary.
By: Darrell St. Claire
Assistant Secretary.

210

IN THE HOUSE OF REPRESENTATIVES, U.S.,
October 18, 1972.

The House of Representatives having proceeded to reconsider the bill (S. 2770) entitled "An Act to amend the Federal Water Pollution Control Act," returned by the President of the United States with his objections to the Senate, in which it originated, it was

Resolved, That the said bill pass, two-thirds of the House of Representatives agreeing to pass the same.

Attest:

W. PAT JENNINGS

Clerk.

By: W. Raymond Colley.

LEGISLATIVE HISTORY:

HOUSE REPORTS: No. 92-911 accompanying H.R. 11896 (Comm. on Public
Works) and No. 92-1465 (Comm. of Conference).
SENATE REPORTS: No. 92-414 (Comm. on Public Works) and No. 92-1236
(Comm. of Conference).
CONGRESSIONAL RECORD:
Vol. 117 (1971): Nov. 2, considered and passed Senate.
Vol. 118 (1972): Mar. 27-29, considered and passed House,
amended, in lieu of H.R. 11896.
Oct. 4, House and Senate agreed to conference
report.
WEEKLY COMPILATION OF PRESIDENTIAL DOCUMENTS:
Vol. 8, No. 43 (1972): Oct. 17, vetoed; Presidential message.
CONGRESSIONAL RECORD:
Vol. 118 (1972): Oct. 18, Senate and House overrode veto.

Summary:

The Federal Water Pollution Control Act was enacted to restore and maintain the chemical, physical, and biological integrity of the Nation's waters. Consistent with the provisions
of the Act, the objectives are to be accomplished through:
eliminating the discharge of pollutants into navigable streams
by 1985; attaining a goal of water quality that will provide
for the protection of fish, shellfish, and wildlife, and provide for recreation in and out of the water by July 1, 1983;
prohibiting the discharge of toxic pollutants in toxic amounts;
providing Federal financial assistance for the construction of
publicly owned waste treatment works; developing and implementing management planning processes so as to assure adequate
control of sources of pollutants in each state; and initiating
a major research and demonstration effort to develop technology necessary to eliminate the discharge of pollutants into
the navigable waters, waters of the contiguous zone, and the
oceans.

Review Questions:

1. What are two specific purposes of the Federal Water
   Pollution Control Act?

2. Identify one economic factor that aided the passage of
   the Federal Water Pollution Control Act?

3. Identify one political factor that aided the passage of
   the Federal Water Pollution Control Act?

4. Identify one social factor that aided the passage of the
   Federal Water Pollution Control Act?

5. What effect has the Federal Water Pollution Control Act
   had on the steel manufacturing industry? the chemical
   manufacturing industry? the petroleum refining industry?

6. What are the strengths of the Federal Water Pollution
   Control Act with regard to current pollution control
   problems?

7. What are the weaknesses of the Federal Water Pollution
   Control Act with regard to current pollution control
   problems?

8. What are the strengths of the Federal Water Pollution
   Control Act with regard to current energy conservation
   measures?

9. What are the weaknesses of the Federal Water Pollution
   Control Act with regard to current energy conservation
   measures?

pesticides

OBJECTIVES:

The student should be able to identify the specific purposes of the Federal Environmental Pesticide Control Act of 1972.

The student should be able to describe those economic, political, and social factors leading to the passage of the Federal Environmental Pesticide Control Act of 1972.

The student should be able to describe the effect the Federal Environmental Pesticide Control Act of 1972 has had on various types of industries.

The student should be able to identify the strengths and weaknesses of the Federal Environmental Pesticide Control Act of 1972 as they relate to current pollution control problems.

INTRODUCTION:

Use of chemicals to control pests has long been practiced in the United States. Decades ago, insects causing harm to agricultural crops were dusted with arsenical compounds or sulphurs, and insects that carried human disease or were regarded as nuisances were fought with sprays of light oils and pyrethrins.

Subsequently, synthetic, organic compounds were developed that effectively killed many insect pests long after the time of application. Other chemicals—the herbicides—regulated the growth of broadleaf plants, still others controlled fungus of many types.

Over the years, several hundred basic chemicals have been created and marketed in thousands of

TOWARD A NEW ENVIRONMENTAL ETHIC, U.S. Government Printing Office: 1971-O — 443-062, pp. 22-23.

formulations to control unwanted insects, plants, fungus growth, soil nematodes, small mammals, and other pests. Not only were agricultural lands treated but homes, gardens, and turf were also covered liberally. The total tonnage of all liquids, granules, and powders rose to the hundreds of thousands and thousands of uses were devised.

The benefits, in terms of increased food production and the control of such killing diseases as malaria and encephalitis all over the world were real and apparent. However, knowledge of the possible side effects of such chemicals entering the environment came more slowly.

By 1944, research had shown that the first chlorinated hydrocarbon compounds such as DDT must be classed as killers of fish. By 1948, their ability to accumulate in fatty tissues became apparent.

It is now known that some of the more persistent compounds are present in the tissues of birds, fish, and other wildlife and man as well. A concentrating effect takes place as one species feeds on another and passes the pesticide from one link to another in the food chain. Hence, certain predator birds, fish, or animals may accumulate levels several thousand times the concentration in their environment.

Man, of course, is at the top of this food chain, and the average American now carries about 12 parts per million of DDT in his fatty tissues. There is no direct evidence that this concentration has a harmful effect on humans. However, there is evidence that concentrated pesticide residues have adverse effects on reproduction, physiology, and behavior in some birds and may threaten the survival of certain desirable species of wildlife.

The oceans are a final accumulation site for many of the persistent chlorinated hydrocarbons. One quarter of the world's entire production of DDT may have been transferred to the sea by now, according to some estimates. Some of its effects on the marine environment have been demonstrated. Marine fish are almost universally contaminated with residues of the persistent pesticides, and declining production of shrimp, crabs, and oysters in certain areas is believed to be directly traceable to pesticide contamination.

Newer types of chemical families include the organophosphates and carbamates. Less persistent but more toxic, these have been responsible for accidental kills of both wildlife and humans. Misuse of various pesticides is implicated in up to 200 human deaths per year and thousands of cases of severe illness.

Ironically, by relying too heavily for pest control on the strategy of chemical extermination, we may well have played into the hands of the insect enemy itself, with its tremendous capacity for adaptation and survival. More and more insect species, including some of those that carry human disease, have developed immunity to the pesticides which had kept them under control. They are thriving again, impervious to such chemical treatment. The answer to our present dilemma will not be found solely in the development of safer more selective chemical formulations, although this is important. It requires, as well, the development of alternative strategies for disease control and crop protection.

Perhaps no environmental problem illustrates more clearly than this one the complex intractions that take place throughout the ecosystem, and the caution that must be exercised to be sure that beneficial changes made by man in one part of the system do not create serious damage in another.

The United States Environmental Protection Agency now exercises the principal regulatory and research functions of the federal government over pesticides under authorities contained in the Federal Insecticide, Fungicide, and Rodenticide Act of 1947, as amended; Section 403(d) of the Federal Food, Drug, and Cosmetic Act, as amended; and other laws.

In brief, the salient points of the Federal program are:

registration and labelling

• Manufacturers must apply to EPA for registration of any insecticide herbicide, fungicide, or similar economic poison that will move in interstate commerce. Such chemicals cannot be approved for sale unless the maker provides scientific evidence that his product is effective for the purpose intended and will not injure human beings, livestock, crops, or wildlife when used as directed. Labelling must show the types and amounts of ingredients, method of application, and precautions to be observed.

EPA continuously reviews registered chemicals in light of developing scientific data to insure continued compliance with requirements for safety and efficacy. EPA inspectors collect samples for laboratory analysis. Field checks are performed periodically to confirm the effectiveness of the compound. Pharmacological tests are made to insure that safety precautions shown on the label are adequate.

• EPA may immediately *suspend* the registration of any pesticide product, thus effectively terminating any further interstate shipments if such action is found to be necessary to prevent "imminent hazard to the public."

• Where "imminent hazard" does not exist, EPA may commence action to terminate a registration by issuing *a notice of cancellation* to become effective after 30 days. If the cancellation is challenged, extensive review procedures are required. During this period the product registration remains valid and interstate marketing may continue.

Cancellations and suspension covering certain uses of DDT, aldrin, dieldrin, and the herbicide 2,4,5-T, initiated by the Secretary of Agriculture prior to the establishment of EPA are now final.

Notices of cancellation covering remaining uses of these chemicals—as well as of mirex, a compound used to control the fire ant—have been issued and the review procedures initiated. Final determinations with regard to continued use of these products may be expected early in 1972.

EPA is also carrying out within the Agency, intensive review of registrations of products, containing benzene hexachloride, lindane, chlordane, endrin, heptachlor, and toxaphene, among others, as well as all products containing mercury, arsenic, or lead.

safe tolerances on foods

• EPA establishes for each registered pesticide a "safe tolerance"—that is, the amount of residue that may be safely permitted on raw food crops—

to protect the public health. The Food and Drug Administration, Department of Health, Education, and Welfare, enforces these tolerances for foods in interstate commerce by regular inspections, and seizure or recall of shipments exceeding the established limits and prosecution of offenders.

research and monitoring

• EPA conducts extensive research on all aspects of pesticides in the environment. At its Primate Research Facility, in Perrine, Florida, it seeks to determine more precisely the effects of pesticides on man. Various community studies are designed to provide a picture of total human exposure to such chemicals and their effects on human health. Studies of the effects on fish and wildlife are carried out, particularly at the Gulf Breeze Laboratory in Florida. Monitoring of soil, air, and water provides knowledge of the levels and pathways of pesticide contamination.

technical assistance

• Technical assistance is given to State agencies to strengthen their pesticide control programs.

Public Law 92-516
92nd Congress, H. R. 10729
October 21, 1972

An Act

To amend the Federal Insecticide, Fungicide, and Rodenticide Act, and for other purposes.

Be it enacted by the Senate and House of Representatives of the United States of America in Congress assembled, That this Act may be cited as the "Federal Environmental Pesticide Control Act of 1972".

Federal Environmental Pesticide Control Act of 1972.

AMENDMENTS TO FEDERAL INSECTICIDE, FUNGICIDE, AND RODENTICIDE ACT

SEC. 2. The Federal Insecticide, Fungicide, and Rodenticide Act (7 U.S.C. 135 et seq.) is amended to read as follows:

61 Stat. 163; 78 Stat. 190.

"SECTION 1. SHORT TITLE AND TABLE OF CONTENTS.

"(a) SHORT TITLE.—This Act may be cited as the 'Federal Insecticide, Fungicide, and Rodenticide Act'.

"(b) TABLE OF CONTENTS.—

FEDERAL ENVIRONMENTAL PESTICIDE CONTROL ACT, October 21, 1972, Public Law 92-516, 86 STAT. 973-999.

218

"SEC. 2. DEFINITIONS.

"For purposes of this Act—

"(a) ACTIVE INGREDIENT.—The term 'active ingredient' means—

"(1) in the case of a pesticide other than a plant regulator, defoliant, or desiccant, an ingredient which will prevent, destroy, repel, or mitigate any pest;

"(2) in the case of a plant regulator, an ingredient which, through physiological action, will accelerate or retard the rate of growth or rate of maturation or otherwise alter the behavior of ornamental or crop plants or the product thereof;

"(3) in the case of a defoliant, an ingredient which will cause the leaves or foliage to drop from a plant; and

"(4) in the case of a desiccant, an ingredient which will artificially accelerate the drying of plant tissue.

"(b) ADMINISTRATOR.—The term 'Administrator' means the Administrator of the Environmental Protection Agency.

"(c) ADULTERATED.—The term 'adulterated' applies to any pesticide if:

"(1) its strength or purity falls below the professed standard of quality as expressed on its labeling under which it is sold;

"(2) any substance has been substituted wholly or in part for the pesticide; or

"(3) any valuable constituent of the pesticide has been wholly or in part abstracted.

"(d) ANIMAL.—The term 'animal' means all vertebrate and invertebrate species, including but not limited to man and other mammals, birds, fish, and shellfish.

"(e) CERTIFIED APPLICATOR, ETC.—

"(1) CERTIFIED APPLICATOR.—The term 'certified applicator' means any individual who is certified under section 4 as authorized to use or supervise the use of any pesticide which is classified for restricted use.

"(2) PRIVATE APPLICATOR.—The term 'private applicator' means a certified applicator who uses or supervises the use of any pesticide which is classified for restricted use for purposes of producing any agricultural commodity on property owned or rented by him or his employer or (if applied without compensation other than trading of personal services between producers of agricultural commodities) on the property of another person.

"(3) COMMERCIAL APPLICATOR.—The term 'commercial applicator' means a certified applicator (whether or not he is a private applicator with respect to some uses) who uses or supervises the use of any pesticide which is classified for restricted use for any purpose or on any property other than as provided by paragraph (2).

"(4) UNDER THE DIRECT SUPERVISION OF A CERTIFIED APPLICATOR.—Unless otherwise prescribed by its labeling, a pesticide shall be considered to be applied under the direct supervision of a certified applicator if it is applied by a competent person acting under the instructions and control of a certified applicator who is available if and when needed, even though such certified applicator is not physically present at the time and place the pesticide is applied.

"(f) DEFOLIANT.—The term 'defoliant' means any substance or mixture of substances intended for causing the leaves or foliage to drop from a plant, with or without causing abscission.

"(g) DESICCANT.—The term 'desiccant' means any substance or mixture of substances intended for artificially accelerating the drying of plant tissue.

"(h) DEVICE.—The term 'device' means any instrument or contrivance (other than a firearm) which is intended for trapping, destroying, repelling, or mitigating any pest or any other form of plant or animal life (other than man and other than bacteria, virus, or other microorganism on or in living man or other living animals); but not including equipment used for the application of pesticides when sold separately therefrom.

"(i) DISTRICT COURT.—The term 'district court' means a United States district court, the District Court of Guam, the District Court of the Virgin Islands, and the highest court of American Samoa.

"(j) ENVIRONMENT.—The term 'environment' includes water, air, land, and all plants and man and other animals living therein, and the interrelationships which exist among these.

"(k) FUNGUS.—The term 'fungus' means any non-chlorophyll-bearing thallophyte (that is, any non-chlorophyll-bearing plant of a lower order than mosses and liverworts), as for example, rust, smut, mildew, mold, yeast, and bacteria, except those on or in living man or other animals and those on or in processed food, beverages, or pharmaceuticals.

"(l) IMMINENT HAZARD.—The term 'imminent hazard' means a situation which exists when the continued use of a pesticide during the time required for cancellation proceeding would be likely to result in unreasonable adverse effects on the environment or will involve unreasonable hazard to the survival of a species declared endangered by the Secretary of the Interior under Public Law 91–135.

83 Stat. 275.
16 USC 668cc-1.

"(m) INERT INGREDIENT.—The term 'inert ingredient' means an ingredient which is not active.

"(n) INGREDIENT STATEMENT.—The term 'ingredient statement' means a statement which contains—

"(1) the name and percentage of each active ingredient, and the total percentage of all inert ingredients, in the pesticide; and

"(2) if the pesticide contains arsenic in any form, a statement of the percentages of total and water soluble arsenic, calculated as elementary arsenic.

"(o) INSECT.—The term 'insect' means any of the numerous small invertebrate animals generally having the body more or less obviously segmented, for the most part belonging to the class insecta, comprising six-legged, usually winged forms, as for example, beetles, bugs, bees, flies, and to other allied classes of arthropods whose members are wingless and usually have more than six legs, as for example, spiders, mites, ticks, centipedes, and wood lice.

"(p) LABEL AND LABELING.—

"(1) LABEL.—The term 'label' means the written, printed, or graphic matter on, or attached to, the pesticide or device or any of its containers or wrappers.

"(2) LABELING.—The term 'labeling' means all labels and all other written, printed, or graphic matter—

"(A) accompanying the pesticide or device at any time; or

"(B) to which reference is made on the label or in literature accompanying the pesticide or device, except to current official publications of the Environmental Protection Agency, the United States Departments of Agriculture and Interior, the Department of Health, Education, and Welfare, State experiment stations, State agricultural colleges, and other similar Federal or State institutions or agencies authorized by law to conduct research in the field of pesticides.

"(q) MISBRANDED.—

"(1) A pesticide is misbranded if—

"(A) its labeling bears any statement, design, or graphic representation relative thereto or to its ingredients which is false or misleading in any particular;

"(B) it is contained in a package or other container or wrapping which does not conform to the standards established by the Administrator pursuant to section 25(c)(3);

"(C) it is an imitation of, or is offered for sale under the name of, another pesticide;

"(D) its label does not bear the registration number assigned under section 7 to each establishment in which it was produced;

"(E) any word, statement, or other information required by or under authority of this Act to appear on the label or labeling is not prominently placed thereon with such conspicuousness (as compared with other words, statements, designs, or graphic matter in the labeling) and in such terms as to render it likely to be read and understood by the ordinary individual under customary conditions of purchase and use;

"(F) the labeling accompanying it does not contain directions for use which are necessary for effecting the purpose for which the product is intended and if complied with, together with any requirements imposed under section 3(d) of this Act, are adequate to protect hea'th and the environment;

"(G) the label does not contain a warning or caution statement which may be necessary and if complied with, together with any requirements imposed under section 3(d) of this Act, is adequate to protect health and the environment.

"(2) A pesticide is misbranded if—

"(A) the label does not bear an ingredient statement on that part of the immediate container (and on the outside container or wrapper of the retail package, if there be one, through which the ingredient statement on the immediate container cannot be clearly read) which is presented or displayed under customary conditions of purchase, except that a pesticide is not misbranded under this subparagraph if:

"(i) the size of form of the immediate container, or the outside container or wrapper of the retail package, makes it impracticable to place the ingredient statement on the part which is presented or displayed under customary conditions of purchase; and

"(ii) the ingredient statement appears prominently on another part of the immediate container, or outside container or wrapper, permitted by the Administrator;

"(B) the labeling does not contain a statement of the use classification under which the product is registered;

"(C) there is not affixed to its container, and to the outside container or wrapper of the retail package, if there be one, through which the required information on the immediate container cannot be clearly read, a label bearing—

"(i) the name and address of the producer, registrant, or person for whom produced;

"(ii) the name, brand, or trademark under which the pesticide is sold;

"(iii) the net weight or measure of the content: *Provided.* That the Administrator may permit reasonable variations; and

"(v) when required by regulation of the Administrator to effectuate the purposes of this Act, the registration number assigned to the pesticide under this Act, and the use classification; and

"(D) the pesticide contains any substance or substances in quantities highly toxic to man, unless the label shall bear, in addition to any other matter required by this Act—

"(i) the skull and crossbones;

"(ii) the word 'poison' prominently in red on a background of distinctly contrasting color; and

"(iii) a statement of a practical treatment (first aid or otherwise) in case of poisoning by the pesticide.

"(r) NEMATODE.—The term 'nematode' means invertebrate animals of the phylum nemathelminthes and class nematoda, that is, unsegmented round worms with elongated, fusiform, or saclike bodies covered with cuticle, and inhabiting soil, water, plants, or plant parts; may also be called nemas or eelworms.

"(s) PERSON.—The term 'person' means any individual, partnership, association, corporation, or any organized group of persons whether incorporated or not.

"(t) PEST.—The term 'pest' means (1) any insect, rodent, nematode, fungus, weed, or (2) any other form of terrestrial or aquatic plant or animal life or virus, bacteria, or other micro-organism (except viruses, bacteria, or other micro-organisms on or in living man or other living animals) which the Administrator declares to be a pest under section 25(c)(1).

"(u) PESTICIDE.—The term 'pesticide' means (1) any substance or mixture of substances intended for preventing, destroying, repelling, or mitigating any pest, and (2) any substance or mixture of substances intended for use as a plant regulator, defoliant, or desiccant.

"(v) PLANT REGULATOR.—The term 'plant regulator' means any substance or mixture of substances intended, through physiological action, for accelerating or retarding the rate of growth or rate of maturation, or for otherwise altering the behavior of plants or the produce thereof, but shall not include substances to the extent that they are intended as plant nutrients, trace elements, nutritional chemicals, plant inoculants, and soil amendments. Also, the term 'plant regulator' shall not be required to include any of such of those nutrient mixtures or soil amendments as are commonly known as vitamin-hormone horticultural products, intended for improvement, maintenance, survival, health, and propagation of plants, and as are not for pest destruction and are nontoxic, nonpoisonous in the undiluted packaged concentration.

"(w) PRODUCER AND PRODUCE.—The term 'producer' means the person who manufactures, prepares, compounds, propagates, or processes any pesticide or device. The term 'produce' means to manufacture, prepare, compound, propagate, or process any pesticide or device.

"(x) PROTECT HEALTH AND THE ENVIRONMENT.—The terms 'protect health and the environment' and 'protection of health and the environment' mean protection against any unreasonable adverse effects on the environment.

"(y) REGISTRANT.—The term 'registrant' means a person who has registered any pesticide pursuant to the provisions of this Act.

"(z) REGISTRATION.—The term 'registration' includes reregistration.

"(aa) STATE.—The term 'State' means a State, the District of Columbia, the Commonwealth of Puerto Rico, the Virgin Islands, Guam, the Trust Territory of the Pacific Islands, and American Samoa.

"(bb) UNREASONABLE ADVERSE EFFECTS ON THE ENVIRONMENT.— The term 'unreasonable adverse effects on the environment' means any unreasonable risk to man or the environment, taking into account the economic, social, and environmental costs and benefits of the use of any pesticide.

"(cc) WEED.—The term 'weed' means any plant which grows where not wanted.

"(dd) ESTABLISHMENT.—The term 'establishment' means any place where a pesticide or device is produced, or held, for distribution or sale.

"SEC. 3. REGISTRATION OF PESTICIDES.

"(a) REQUIREMENT.—Except as otherwise provided by this Act, no person in any State may distribute, sell, offer for sale, hold for sale, ship, deliver for shipment, or receive and (having so received) deliver or offer to deliver, to any person any pesticide which is not registered with the Administrator.

"(b) EXEMPTIONS.—A pesticide which is not registered with the Administrator may be transferred if—

"(1) the transfer is from one registered establishment to another registered establishment operated by the same producer solely for packaging at the second establishment or for use as a constituent part of another pesticide produced at the second establishment; or

"(2) the transfer is pursuant to and in accordance with the requirements of an experimental use permit.

"(c) PROCEDURE FOR REGISTRATION.—

"(1) STATEMENT REQUIRED.—Each applicant for registration of a pesticide shall file with the Administrator a statement which includes—

"(A) the name and address of the applicant and of any other person whose name will appear on the labeling;

"(B) the name of the pesticide;

"(C) a complete copy of the labeling of the pesticide, a statement of all claims to be made for it, and any directions for its use;

"(D) if requested by the Administrator, a full description of the tests made and the results thereof upon which the

224

claims are based, except that data submitted in support of an application shall not, without permission of the applicant, be considered by the Administrator in support of any other application for registration unless such other applicant shall have first offered to pay reasonable compensation for producing the test data to be relied upon and such data is not protected from disclosure by section 10(b). If the parties cannot agree on the amount and method of payment, the Administrator shall make such determination and may fix such other terms and conditions as may be reasonable under the circumstances. The Administrator's determination shall be made on the record after notice and opportunity for hearing. If the owner of the test data does not agree with said determination, he may, within thirty days, take an appeal to the federal district court for the district in which he resides with respect to either the amount of the payment or the terms of payment, or both. In no event shall the amount of payment determined by the court be less than that determined by the Administrator;

"(E) the complete formula of the pesticide; and

"(F) a request that the pesticide be classified for general use, for restricted use, or for both.

"(2) DATA IN SUPPORT OF REGISTRATION.—The Administrator shall publish guidelines specifying the kinds of information which will be required to support the registration of a pesticide and shall revise such guidelines from time to time. If thereafter he requires any additional kind of information he shall permit sufficient time for applicants to obtain such additional information. Except as provided by subsection (c)(1)(D) of this section and section 10, within 30 days after the Administrator registers a pesticide under this Act he shall make available to the public the data called for in the registration statement together with such other scientific information as he deems relevant to his decision.

"(3) TIME FOR ACTING WITH RESPECT TO APPLICATION.—The Administrator shall review the data after receipt of the application and shall, as expeditiously as possible, either register the pesticide in accordance with paragraph (5), or notify the applicant of his determination that it does not comply with the provisions of the Act in accordance with paragraph (6).

Publication in Federal Register.

"(4) NOTICE OF APPLICATION.—The Administrator shall publish in the Federal Register, promptly after receipt of the statement and other data required pursuant to paragraphs (1) and (2), a notice of each application for registration of any pesticide if it contains any new active ingredient or if it would entail a changed use pattern. The notice shall provide for a period of 30 days in which any Federal agency or any other interested person may comment.

"(5) APPROVAL OF REGISTRATION.—The Administrator shall register a pesticide if he determines that, when considered with any restrictions imposed under subsection (d)—

"(A) its composition is such as to warrant the proposed claims for it;

"(B) its labeling and other material required to be submitted comply with the requirements of this Act;

"(C) it will perform its intended function without unreasonable adverse effects on the environment; and

225

"(D) when used in accordance with widespread and commonly recognized practice it will not generally cause unreasonable adverse effects on the environment.

The Administrator shall not make any lack of essentiality a criterion for denying registration of any pesticide. Where two pesticides meet the requirements of this paragraph, one should not be registered in preference to the other.

"(6) DENIAL OF REGISTRATION.—If the Administrator determines that the requirements of paragraph (5) for registration are not satisfied, he shall notify the applicant for registration of his determination and of his reasons (including the factual basis) therefor, and that, unless the applicant corrects the conditions and notifies the Administrator thereof during the 30-day period beginning with the day after the date on which the applicant receives the notice, the Administrator may refuse to register the pesticide. Whenever the Administrator refuses to register a pesticide, he shall notify the applicant of his decision and of his reasons (including the factual basis) therefor. The Administrator shall promptly publish in the Federal Register notice of such denial of registration and the reasons therefor. Upon such notification, the applicant for registration or other interested person with the concurrence of the applicant shall have the same remedies as provided for in section 6.

Publication in Federal Register.

"(d) CLASSIFICATION OF PESTICIDES.—

"(1) CLASSIFICATION FOR GENERAL USE, RESTRICTED USE, OR BOTH.—

"(A) As a part of the registration of a pesticide the Administrator shall classify it as being for general use or for restricted use, provided that if the Administrator determines that some of the uses for which the pesticide is registered should be for general use and that other uses for which it is registered should be for restricted use, he shall classify it for both general use and restricted use. If some of the uses of the pesticide are classified for general use and other uses are classified for restricted use, the directions relating to its general uses shall be clearly separated and distinguished from those directions relating to its restricted uses: *Provided, however*, That the Administrator may require that its packaging and labeling for restricted uses shall be clearly distinguishable from its packaging and labeling for general uses.

"(B) If the Administrator determines that the pesticide, when applied in accordance with its directions for use, warnings and cautions and for the uses for which it is registered, or for one or more of such uses, or in accordance with a widespread and commonly recognized practice, will not generally cause unreasonable adverse effects on the environment, he will classify the pesticide, or the particular use or uses of the pesticide to which the determination applies, for general use.

"(C) If the Administrator determines that the pesticide, when applied in accordance with its directions for use, warnings and cautions and for the uses for which it is registered, or for one or more of such uses, or in accordance with a widespread and commonly recognized practice, may generally cause, without additional regulatory restrictions, unreasonable adverse effects on the environment, including injury to the applicator, he shall classify the pesticide, or the particular use or uses to which the determination applies, for restricted use:

"(i) If the Administrator classifies a pesticide, or one or more uses of such pesticide, for restricted use because of a determination that the acute dermal or inhalation toxicity of the pesticide presents a hazard to the applicator or other persons, the pesticide shall be applied for any use to which the restricted classification applies only by or under the direct supervision of a certified applicator.

"(ii) If the Administrator classifies a pesticide, or one or more uses of such pesticide, for restricted use because of a determination that its use without additional regulatory restriction may cause unreasonable adverse effects on the environment, the pesticide shall be applied for any use to which the determination applies only by or under the direct supervision of a certified applicator, or subject to such other restrictions as the Administrator may provide by regulation. Any such regulation shall be reviewable in the appropriate court of appeals upon petition of a person adversely affected filed within 60 days of the publication of the regulation in final form.

Publication in Federal Register. "(2) CHANGE IN CLASSIFICATION.—If the Administrator determines that a change in the classification of any use of a pesticide from general use to restricted use is necessary to prevent unreasonable adverse effects on the environment, he shall notify the registrant of such pesticide of such determination at least 30 days before making the change and shall publish the proposed change in the Federal Register. The registrant, or other interested person with the concurrence of the registrant, may seek relief from such determination under section 6(b).

"(e) PRODUCTS WITH SAME FORMULATION AND CLAIMS.—Products which have the same formulation, are manufactured by the same person, the labeling of which contains the same claims, and the labels of which bear a designation identifying the product as the same pesticide may be registered as a single pesticide; and additional names and labels shall be added to the registration by supplemental statements.

"(f) MISCELLANEOUS.—

"(1) EFFECT OF CHANGE OF LABELING OR FORMULATION.—If the labeling or formulation for a pesticide is changed, the registration shall be amended to reflect such change if the Administrator determines that the change will not violate any provision of this Act.

"(2) REGISTRATION NOT A DEFENSE.—In no event shall registration of an article be construed as a defense for the commission of any offense under this Act: *Provided*, That as long as no cancellation proceedings are in effect registration of a pesticide shall be prima facie evidence that the pesticide, its labeling and packaging comply with the registration provisions of the Act.

"(3) AUTHORITY TO CONSULT OTHER FEDERAL AGENCIES.—In connection with consideration of any registration or application for registration under this section, the Administrator may consult with any other Federal agency.

227

"SEC. 4. USE OF RESTRICTED USE PESTICIDES; CERTIFIED APPLICATORS.

"(a) CERTIFICATION PROCEDURE.—

"(1) FEDERAL CERTIFICATION.—Subject to paragraph (2), the Administrator shall prescribe standards for the certification of applicators of pesticides. Such standards shall provide that to be certified, an individual must be determined to be competent with respect to the use and handling of pesticides, or to the use and handling of the pesticide or class of pesticides covered by such individual's certification. *Standards.*

"(2) STATE CERTIFICATION.—If any State, at any time, desires to certify applicators of pesticides, the Governor of such State shall submit a State plan for such purpose. The Administrator shall approve the plan submitted by any State, or any modification thereof, if such plan in his judgment—

"(A) designates a State agency as the agency responsible for administering the plan throughout the State;

"(B) contains satisfactory assurances that such agency has or will have the legal authority and qualified personnel necessary to carry out the plan;

"(C) gives satisfactory assurances that the State will devote adequate funds to the administration of the plan;

"(D) provides that the State agency will make such reports to the Administrator in such form and containing such information as the Administrator may from time to time require; and

"(E) contains satisfactory assurances that State standards for the certification of applicators of pesticides conform with those standards prescribed by the Administrator under paragraph (1).

Any State certification program under this section shall be maintained in accordance with the State plan approved under this section.

"(b) STATE PLANS.—If the Administrator rejects a plan submitted under this paragraph, he shall afford the State submitting the plan due notice and opportunity for hearing before so doing. If the Administrator approves a plan submitted under this paragraph, then such State shall certify applicators of pesticides with respect to such State. Whenever the Administrator determines that a State is not administering the certification program in accordance with the plan approved under this section, he shall so notify the State and provide for a hearing at the request of the State, and, if appropriate corrective action is not taken within a reasonable time, not to exceed ninety days, the Administrator shall withdraw approval of such plan. *Hearing.*

"SEC. 5. EXPERIMENTAL USE PERMITS.

"(a) ISSUANCE.—Any person may apply to the Administrator for an experimental use permit for a pesticide. The Administrator may issue an experimental use permit if he determines that the applicant needs such permit in order to accumulate information necessary to register a pesticide under section 3. An application for an experimental use permit may be filed at the time of or before or after an application for registration is filed.

"(b) TEMPORARY TOLERANCE LEVEL.—If the Administrator determines that the use of a pesticide may reasonably be expected to result in any residue on or in food or feed, he may establish a temporary tolerance level for the residue of the pesticide before issuing the experimental use permit.

228

"(c) USE UNDER PERMIT.—Use of a pesticide under an experimental use permit shall be under the supervision of the Administrator, and shall be subject to such terms and conditions and be for such period of time as the Administrator may prescribe in the permit.

"(d) STUDIES.—When any experimental use permit is issued for a pesticide containing any chemical or combination of chemicals which has not been included in any previously registered pesticide, the Administrator may specify that studies be conducted to detect whether the use of the pesticide under the permit may cause unreasonable adverse effects on the environment. All results of such studies shall be reported to the Administrator before such pesticide may be registered under section 3.

"(e) REVOCATION.—The Administrator may revoke any experimental use permit, at any time, if he finds that its terms or conditions are being violated, or that its terms and conditions are inadequate to avoid unreasonable adverse effects on the environment.

"(f) STATE ISSUANCE OF PERMITS.—Notwithstanding the foregoing provisions of this section, the Administrator may, under such terms and conditions as he may by regulations prescribe, authorize any State to issue an experimental use permit for a pesticide. All provisions of section 4 relating to State plans shall apply with equal force to a State plan for the issuance of experimental use permits under this section.

"SEC. 6. ADMINISTRATIVE REVIEW; SUSPENSION.

"(a) CANCELLATION AFTER FIVE YEARS—

"(1) PROCEDURE.—The Administrator shall cancel the registration of any pesticide at the end of the five-year period which begins on the date of its registration (or at the end of any five-year period thereafter) unless the registrant, or other interested person with the concurrence of the registrant, before the end of such period, requests in accordance with regulations prescribed by the Administrator that the registration be continued in effect: *Provided*, That the Administrator may permit the continued sale and use of existing stocks of a pesticide whose registration is canceled under this subsection or subsection (b) to such extent, under such conditions, and for such uses as he may specify if he determines that such sale or use is not inconsistent with the purposes of this Act and will not have unreasonable adverse effects on the environment. The Administrator shall publish in the Federal Register, at least 30 days prior to the expiration of such five-year period, notice that the registration will be canceled if the registrant or other interested person with the concurrence of the registrant does not request that the registration be continued in effect.

Publication in Federal Register.

"(2) INFORMATION.—If at any time after the registration of a pesticide the registrant has additional factual information regarding unreasonable adverse effects on the environment of the pesticide, he shall submit such information to the Administrator.

"(b) CANCELLATION AND CHANGE IN CLASSIFICATION.—If it appears to the Administrator that a pesticide or its labeling or other material required to be submitted does not comply with the provisions of this Act or, when used in accordance with widespread and commonly recognized practice, generally causes unreasonable adverse effects on the environment, the Administrator may issue a notice of his intent either—

"(1) to cancel its registration or to change its classification together with the reasons (including the factual basis) for his action, or

"(2) to hold a hearing to determine whether or not its registration should be canceled or its classification changed. Hearing.

Such notice shall be sent to the registrant and made public. The proposed action shall become final and effective at the end of 30 days from receipt by the registrant, or publication, of a notice issued under paragraph (1), whichever occurs later, unless within that time either (i) the registrant makes the necessary corrections, if possible, or (ii) a request for a hearing is made by a person adversely affected by the notice. In the event a hearing is held pursuant to such a request or to the Administrator's determination under paragraph (2), a decision pertaining to registration or classification issued after completion of such hearing shall be final.

"(c) Suspension.—

"(1) Order.—If the Administrator determines that action is necessary to prevent an imminent hazard during the time required for cancellation or change in classification proceedings, he may, by order, suspend the registration of the pesticide immediately. No order of suspension may be issued unless the Administrator has issued or at the same time issues notice of his intention to cancel the registration or change the classification of the pesticide.

"Except as provided in paragraph (3), the Administrator shall notify the registrant prior to issuing any suspension order. Such notice shall include findings pertaining to the question of 'imminent hazard'. The registrant shall then have an opportunity, in accordance with the provisions of paragraph (2), for an expedited hearing before the Agency on the question of whether an imminent hazard exists.

"(2) Expedite hearing.—If no request for a hearing is submitted to the Agency within five days of the registrant's receipt of the notification provided for by paragraph (1), the suspension order may be issued and shall take effect and shall not be reviewable by a court. If a hearing is requested, it shall commence within five days of the receipt of the request for such hearing unless the registrant and the Agency agree that it shall commence at a later time. The hearing shall be held in accordance with the provisions of subchapter II of title 5 of the United States Code, except that the presiding officer need not be a certified hearing examiner. The presiding officer shall have ten days from the conclusion of the presentation of evidence to submit recommended findings and conclusions to the Administrator, who shall then have seven days to render a final order on the issue of suspension. 80 Stat. 381; 81 Stat. 54. 5 USC 551.

"(3) Emergency order.—Whenever the Administrator determines that an emergency exists that does not permit him to hold a hearing before suspending, he may issue a suspension order in advance of notification to the registrant. In that case, paragraph (2) shall apply except that (i) the order of suspension shall be in effect pending the expeditious completion of the remedies provided by that paragraph and the issuance of a final order on suspension, and (ii) no party other than the registrant and the Agency shall participate except that any person adversely affected may file briefs within the time allotted by the Agency's rules. Any person so filing briefs shall be considered a party to such proceeding for the purposes of section 16(b).

"(4) Judicial review.—A final order on the question of suspension following a hearing shall be reviewable in accordance with Section 16 of this Act, notwithstanding the fact that any related cancellation proceedings have not been completed. Petitions to review orders on the issue of suspension shall be advanced on the docket of the courts of appeals. Any order of suspension entered prior to a hearing before the Administrator shall be subject to immediate review in an action by the registrant or other interested person with the concurrence of the registrant in an appropriate district court, solely to determine whether the order of suspension was arbitrary, capricious or an abuse of discretion, or whether the order was issued in accordance with the procedures established by law. The effect of any order of the court will be only to stay the effectiveness of the suspension order, pending the Administrator's final decision with respect to cancellation or change in classification. This action may be maintained simultaneously with any administrative review proceeding under this section. The commencement of proceedings under this paragraph shall not operate as a stay of order, unless ordered by the court.

"(d) Public Hearings and Scientific Review.—In the event a hearing is requested pursuant to subsection (b) or determined upon by the Administrator pursuant to subsection (b), such hearing shall be held after due notice for the purpose of receiving evidence relevant and material to the issues raised by the objections filed by the applicant or other interested parties, or to the issues stated by the Administrator, if the hearing is called by the Administrator rather than by the filing of objections. Upon a showing of relevance and reasonable scope of evidence sought by any party to a public hearing, the Hearing Examiner shall issue a subpena to compel testimony or production of documents from any person. The Hearing Examiner shall be guided by the principles of the Federal Rules of Civil Procedure in making any order for the protection of the witness or the content of documents produced and shall order the payment of reasonable fees and expenses as a condition to requiring testimony of the witness. On contest, the subpena may be enforced by an appropriate United States district court in accordance with the principles stated herein. Upon the request of any party to a public hearing and when in the Hearing Examiner's judgment it is necessary or desirable, the Hearing Examiner shall at any time before the hearing record is closed refer to a Committee of the National Academy of Sciences the relevant questions of scientific fact involved in the public hearing. No member of any committee of the National Academy of Sciences established to carry out the functions of this section shall have a financial or other conflict of interest with respect to any matter considered by such committee. The Committee of the National Academy of Sciences shall report in writing to the Hearing Examiner within 60 days after such referral on these questions of scientific fact. The report shall be made public and shall be considered as part of the hearing record. The Administrator shall enter into appropriate arrangements with the National Academy of Sciences to assure an objective and competent scientific review of the questions presented to Committees of the Academy and to provide such other scientific advisory services as may be required by the Administrator for carrying out the purposes of this Act. As soon as practicable after completion of the hearing (including the report of the Academy) but not later than 90 days thereafter, the Administrator shall evaluate the data and reports before him and issue an order either revoking his notice of intention issued pursuant to

Subpena.

28 USC app.

Report.

this section, or shall issue an order either canceling the registration, changing the classification, denying the registration, or requiring modification of the labeling or packaging of the article. Such order shall be based only on substantial evidence of record of such hearing and shall set forth detailed findings of fact upon which the order is based.

"(e) JUDICIAL REVIEW.—Final orders of the Administrator under this section shall be subject to judicial review pursuant to section 16.

"SEC. 7. REGISTRATION OF ESTABLISHMENTS.

"(a) REQUIREMENT.—No person shall produce any pesticide subject to this Act in any State unless the establishment in which it is produced is registered with the Administrator. The application for registration of any establishment shall include the name and address of the establishment and of the producer who operates such establishment.

"(b) REGISTRATION.—Whenever the Administrator receives an application under subsection (a), he shall register the establishment and assign it an establishment number.

"(c) INFORMATION REQUIRED.—

"(1) Any producer operating an establishment registered under this section shall inform the Administrator within 30 days after it is registered of the types and amounts of pesticides—

"(A) which he is currently producing;

"(B) which he has produced during the past year; and

"(C) which he has sold or distributed during the past year. The information required by this paragraph shall be kept current and submitted to the Administrator annually as required under such regulations as the Administrator may prescribe.

"(2) Any such producer shall, upon the request of the Administrator for the purpose of issuing a stop sale order pursuant to section 13, inform him of the name and address of any recipient of any pesticide produced in any registered establishment which he operates.

"(d) CONFIDENTIAL RECORDS AND INFORMATION.—Any information submitted to the Administrator pursuant to subsection (c) shall be considered confidential and shall be subject to the provisions of section 10.

"SEC. 8. BOOKS AND RECORDS.

"(a) REQUIREMENTS.—The Administrator may prescribe regulations requiring producers to maintain such records with respect to their operations and the pesticides and devices produced as he determines are necessary for the effective enforcement of this Act. No records required under this subsection shall extend to financial data, sales data other than shipment data, pricing data, personnel data, and research data (other than data relating to registered pesticides or to a pesticide for which an application for registration has been filed).

Regulations.

"(b) INSPECTION.—For the purposes of enforcing the provisions of this Act, any producer, distributor, carrier, dealer, or any other person who sells or offers for sale, delivers or offers for delivery any pesticide or device subject to this Act, shall, upon request of any officer or employee of the Environmental Protection Agency or of any State or political subdivision, duly designated by the Administrator, furnish or permit such person at all reasonable times to have access to, and to copy: (1) all records showing the delivery, movement, or holding of such pesticide or device, including the quantity, the date of shipment and receipt, and the name of the consignor and consignee; or (2) in the event of the inability of any person to produce records containing such

232

information, all other records and information relating to such delivery, movement, or holding of the pesticide or device. Any inspection with respect to any records and information referred to in this subsection shall not extend to financial data, sales data other than shipment data, pricing data, personnel data, and research data (other than data relating to registered pesticides or to a pesticide for which an application for registration has been filed).

"SEC. 9. INSPECTION OF ESTABLISHMENTS, ETC.

"(a) IN GENERAL.—For purposes of enforcing the provisions of this Act, officers or employees duly designated by the Administrator are authorized to enter at reasonable times, any establishment or other place where pesticides or devices are held for distribution or sale for the purpose of inspecting and obtaining samples of any pesticides or devices, packaged, labeled, and released for shipment, and samples of any containers or labeling for such pesticides or devices.

Before undertaking such inspection, the officers or employees must present to the owner, operator, or agent in charge of the establishment or other place where pesticides or devices are held for distribution or sale, appropriate credentials and a written statement as to the reason for the inspection, including a statement as to whether a violation of the law is suspected. If no violation is suspected, an alternate and sufficient reason shall be given in writing. Each such inspection shall be commenced and completed with reasonable promptness. If the officer or employee obtains any samples, prior to leaving the premises, he shall give to the owner, operator, or agent in charge a receipt describing the samples obtained and, if requested, a portion of each such sample equal in volume or weight to the portion retained. If an analysis is made of such samples, a copy of the results of such analysis shall be furnished promptly to the owner, operator, or agent in charge.

"(b) WARRANTS.—For purposes of enforcing the provisions of this Act and upon a showing to an officer or court of competent jurisdiction that there is reason to believe that the provisions of this Act have been violated, officers or employees duly designated by the Administrator are empowered to obtain and to execute warrants authorizing—

"(1) entry for the purpose of this section;

"(2) inspection and reproduction of all records showing the quantity, date of shipment, and the name of consignor and consignee of any pesticide or device found in the establishment which is adulterated, misbranded, not registered (in the case of a pesticide) or otherwise in violation of this Act and in the event of the inability of any person to produce records containing such information, all other records and information relating to such delivery, movement, or holding of the pesticide or device; and

"(3) the seizure of any pesticide or device which is in violation of this Act.

"(c) ENFORCEMENT.—

"(1) CERTIFICATION OF FACTS TO ATTORNEY GENERAL.—The examination of pesticides or devices shall be made in the Environmental Protection Agency or elsewhere as the Administrator may designate for the purpose of determining from such examinations whether they comply with the requirements of this Act. If it shall appear from any such examination that they fail to comply with the requirements of this Act, the Administrator shall cause notice to be given to the person against whom criminal or civil proceedings are contemplated. Any person so notified shall be given an opportunity to present his views, either orally or in writing, with regard to such contemplated proceedings, and if in the opinion of the Administrator it appears that the provisions of this Act have been violated by such person, then the Administrator shall certify the facts to the Attorney General, with a copy of the results of the analysis or the examination of such pesticide for the institution of a criminal proceeding pursuant to section 14(b) or a civil proceeding under section 14(a), when the Administrator determines that such action will be sufficient to effectuate the purposes of this Act.

"(2) NOTICE NOT REQUIRED.—The notice of contemplated proceedings and opportunity to present views set forth in this subsection are not prerequisites to the institution of any proceeding by the Attorney General.

"(3) WARNING NOTICES.—Nothing in this Act shall be construed as requiring the Administrator to institute proceedings for prosecution of minor violations of this Act whenever he believes that the public interest will be adequately served by a suitable written notice of warning.

"SEC. 10. PROTECTION OF TRADE SECRETS AND OTHER INFORMATION.

"(a) IN GENERAL.—In submitting data required by this Act, the applicant may (1) clearly mark any portions thereof which in his opinion are trade secrets or commercial or financial information and (2) submit such marked material separately from other material required to be submitted under this Act.

"(b) DISCLOSURE.—Notwithstanding any other provision of this Act, the Administrator shall not make public information which in his judgment contains or relates to trade secrets or commercial or financial information obtained from a person and privileged or confidential, except that, when necessary to carry out the provisions of this Act, information relating to formulas of products acquired by authorization of this Act may be revealed to any Federal agency consulted and may be revealed at a public hearing or in findings of fact issued by the Administrator.

"(c) DISPUTES.—If the Administrator proposes to release for inspection information which the applicant or registrant believes to be protected from disclosure under subsection (b), he shall notify the applicant or registrant, in writing, by certified mail. The Administrator shall not thereafter make available for inspection such data until thirty days after receipt of the notice by the applicant or registrant. During this period, the applicant or registrant may institute an action in an appropriate district court for a declaratory judgment as to whether such information is subject to protection under subsection (b).

"SEC. 11. STANDARDS APPLICABLE TO PESTICIDE APPLICATORS.

"(a) IN GENERAL.—No regulations prescribed by the Administrator for carrying out the provisions of this Act shall require any private applicator to maintain any records or file any reports or other documents.

"(b) SEPARATE STANDARDS.—When establishing or approving standards for licensing or certification, the Administrator shall establish separate standards for commercial and private applicators.

"SEC. 12. UNLAWFUL ACTS.

"(a) IN GENERAL.—

"(1) Except as provided by subsection (b), it shall be unlawful for any person in any State to distribute, sell, offer for sale, hold for sale, ship, deliver for shipment, or receive and (having so received) deliver or offer to deliver, to any person—

"(A) any pesticide which is not registered under section 3, except as provided by section 6(a)(1);

"(B) any registered pesticide if any claims made for it as a part of its distribution or sale substantially differ from any claims made for it as a part of the statement required in connection with its registration under section 3;

"(C) any registered pesticide the composition of which differs at the time of its distribution or sale from its composition as described in the statement required in connection with its registration under section 3;

"(D) any pesticide which has not been colored or discolored pursuant to the provisions of section 25(c)(5);

"(E) any pesticide which is adulterated or misbranded; or

"(F) any device which is misbranded.

"(2) It shall be unlawful for any person—

"(A) to detach, alter, deface, or destroy, in whole or in part, any labeling required under this Act;

"(B) to refuse to keep any records required pursuant to section 8, or to refuse to allow the inspection of any records or establishment pursuant to section 8 or 9, or to refuse to allow an officer or employee of the Environmental Protection Agency to take a sample of any pesticide pursuant to section 9;

"(C) to give a guaranty or undertaking provided for in subsection (b) which is false in any particular, except that a person who receives and relies upon a guaranty authorized under subsection (b) may give a guaranty to the same effect, which guaranty shall contain, in addition to his own name and address, the name and address of the person residing in the United States from whom he received the guaranty or undertaking;

"(D) to use for his own advantage or to reveal, other than to the Administrator, or officials or employees of the Environmental Protection Agency or other Federal executive agencies, or to the courts, or to physicians, pharmacists, and other qualified persons, needing such information for the performance of their duties, in accordance with such directions as the Administrator may prescribe, any information acquired by authority of this Act which is confidential under this Act;

"(E) who is a registrant, wholesaler, dealer, retailer, or other distributor to advertise a product registered under this Act for restricted use without giving the classification of the product assigned to it under section 3;

"(F) to make available for use, or to use, any registered pesticide classified for restricted use for some or all purposes other than in accordance with section 3(d) and any regulations thereunder;

"(G) to use any registered pesticide in a manner inconsistent with its labeling;

"(H) to use any pesticide which is under an experimental use permit contrary to the provisions of such permit;

"(I) to violate any order issued under section 13;

"(J) to violate any suspension order issued under section 6;

"(K) to violate any cancellation of registration of a pesticide under section 6, except as provided by section 6(a)(1);

"(L) who is a producer to violate any of the provisions of section 7;

"(M) to knowingly falsify all or part of any application for registration, application for experimental use permit, any information submitted to the Administrator pursuant to section 7, any records required to be maintained pursuant to section 8, any report filed under this Act, or any information marked as confidential and submitted to the Administrator under any provision of this Act;

"(N) who is a registrant, wholesaler, dealer, retailer, or other distributor to fail to file reports required by this Act;

"(O) to add any substance to, or take any substance from, any pesticide in a manner that may defeat the purpose of this Act; or

"(P) to use any pesticide in tests on human beings unless such human beings (i) are fully informed of the nature and purposes of the test and of any physical and mental health consequences which are reasonably foreseeable therefrom, and (ii) freely volunteer to participate in the test.

"(b) EXEMPTIONS.—The penalties provided for a violation of paragraph (1) of subsection (a) shall not apply to—

"(1) any person who establishes a guaranty signed by, and containing the name and address of, the registrant or person residing in the United States from whom he purchased or received in good faith the pesticide in the same unbroken package, to the effect that the pesticide was lawfully registered at the time of sale and delivery to him, and that it complies with the other requirements of this Act, and in such case the guarantor shall be subject to the penalties which would otherwise attach to the person holding the guaranty under the provisions of this Act;

"(2) any carrier while lawfully shipping, transporting, or delivering for shipment any pesticide or device, if such carrier upon request of any officer or employee duly designated by the Administrator shall permit such officer or employee to copy all of its records concerning such pesticide or device;

"(3) any public official while engaged in the performance of his official duties;

"(4) any person using or possessing any pesticide as provided by an experimental use permit in effect with respect to such pesticide and such use or possession; or

"(5) any person who ships a substance or mixture of substances being put through tests in which the purpose is only to determine its value for pesticide purposes or to determine its toxicity or other properties and from which the user does not expect to receive any benefit in pest control from its use.

"SEC. 13. STOP SALE, USE, REMOVAL, AND SEIZURE.

"(a) STOP SALE, ETC., ORDERS.—Whenever any pesticide or device is found by the Administrator in any State and there is reason to believe on the basis of inspection or tests that such pesticide or device is in violation of any of the provisions of this Act, or that such pesticide or device has been or is intended to be distributed or sold in violation of any such provisions, or when the registration of the pesticide has been canceled by a final order or has been suspended, the Administrator may issue a written or printed 'stop sale, use, or removal' order to any person who owns, controls, or has custody of such pesticide or device, and after receipt of such order no person shall sell, use, or remove the pesticide or device described in the order except in accordance with the provisions of the order.

"(b) SEIZURE.—Any pesticide or device that is being transported or, having been transported, remains unsold or in original unbroken packages, or that is sold or offered for sale in any State, or that is imported from a foreign country, shall be liable to be proceeded against in any district court in the district where it is found and seized for confiscation by a process in rem for condemnation if—

"(1) in the case of a pesticide—

"(A) it is adulterated or misbranded;

"(B) it is not registered pursuant to the provisions of section 3;

"(C) its labeling fails to bear the information required by this Act;

"(D) it is not colored or discolored and such coloring or discoloring is required under this Act; or

"(E) any of the claims made for it or any of the directions for its use differ in substance from the representations made in connection with its registration;

"(2) in the case of a device, it is misbranded; or

"(3) in the case of a pesticide or device, when used in accordance with the requirements imposed under this Act and as directed by the labeling, it nevertheless causes unreasonable adverse effects on the environment. In the case of a plant regulator, defoliant, or desiccant, used in accordance with the label claims and recommendations, physical or physiological effects on plants or parts thereof shall not be deemed to be injury, when such effects are the purpose for which the plant regulator, defoliant. or desiccant was applied.

"(c) DISPOSITION AFTER CONDEMNATION.—If the pesticide or device is condemned it shall, after entry of the decree, be disposed of by destruction or sale as the court may direct and the proceeds, if sold, less the court costs, shall be paid into the Treasury of the United States, but the pesticide or device shall not be sold contrary to the provisions of this Act or the laws of the jurisdiction in which it is sold: *Provided*, That upon payment of the costs of the condemnation proceedings and the execution and delivery of a good and sufficient bond conditioned that the pesticide or device shall not be sold or otherwise disposed of contrary to the provisions of the Act or the laws of any jurisdiction in which sold, the court may direct that such pesticide or device be delivered to the owner thereof. The proceedings of such condemnation cases shall confrom, as near as may be to the proceedings in admiralty, except that either party may demand trial by jury of any issue of fact joined in any case, and all such proceedings shall be at the suit of and in the name of the United States.

"(d) COURT COSTS, ETC.—When a decree of condemnation is entered against the pesticide or device, court costs and fees, storage, and other proper expenses shall be awarded against the person, if any, intervening as claimant of the pesticide or device.

"SEC. 14. PENALTIES.

"(a) CIVIL PENALTIES.—

"(1) IN GENERAL.—Any registrant, commercial applicator, wholesaler, dealer, retailer, or other distributor who violates any provision of this Act may be assessed a civil penalty by the Administrator of not more than $5,000 for each offense.

237

"(2) Private Applicator.—Any private applicator or other person not included in paragraph (1) who violates any provision of this Act subsequent to receiving a written warning from the Administrator or following a citation for a prior violation, may be assessed a civil penalty by the Administrator of not more than $1,000 for each offense.

"(3) Hearing.—No civil penalty shall be assessed unless the person charged shall have been given notice and opportunity for a hearing on such charge in the county, parish, or incorporated city of the residence of the person charged. In determining the amount of the penalty the Administrator shall consider the appropriateness of such penalty to the size of the business of the person charged, the effect on the person's ability to continue in business, and the gravity of the violation.

"(4) References to Attorney General.—In case of inability to collect such civil penalty or failure of any person to pay all, or such portion of such civil penalty as the Administrator may determine, the Administrator shall refer the matter to the Attorney General, who shall recover such amount by action in the appropriate United States district court.

"(b) Criminal Penalties.—

"(1) In general.—Any registrant, commercial applicator, wholesaler, dealer, retailer, or other distributor who knowingly violates any provision of this Act shall be guilty of a misdemeanor and shall on conviction be fined not more than $25,000, or imprisoned for not more than one year, or both.

"(2) Private applicator.—Any private applicator or other person not included in paragraph (1) who knowingly violates any provision of this Act shall be guilty of a misdemeanor and shall on conviction be fined not more than $1,000, or imprisoned for not more than 30 days, or both.

"(3) Disclosure of information.—Any person, who, with intent to defraud, uses or reveals information relative to formulas of products acquired under the authority of section 3, shall be fined not more than $10,000, or imprisoned for not more than three years, or both.

"(4) Acts of officers, agents, etc.—When construing and enforcing the provisions of this Act, the act, omission, or failure of any officer, agent, or other person acting for or employed by any person shall in every case be also deemed to be the act, omission, or failure of such person as well as that of the person employed.

"SEC. 15. INDEMNITIES.

"(a) Requirement.—If—

"(1) the Administrator notifies a registrant that he has suspended the registration of a pesticide because such action is necessary to prevent an imminent hazard;

"(2) the registration of the pesticide is canceled as a result of a final determination that the use of such pesticide will create an imminent hazard; and

"(3) any person who owned any quantity of such pesticide immediately before the notice to the registrant under paragraph (1) suffered losses by reason of suspension or cancellation of the registration,

the Administrator shall make an indemnity payment to such person, unless the Administrator finds that such person (i) had knowledge of facts which, in themselves, would have shown that such pesticide did not meet the requirements of section 3(c)(5) for registration, and (ii) continued thereafter to produce such pesticide without giving timely notice of such facts to the Administrator.

"(b) AMOUNT OF PAYMENT.—

"(1) IN GENERAL.—The amount of the idemnity payment under subsection (a) to any person shall be determined on the basis of the cost of the pesticide owned by such person immediately before the notice to the registrant referred to in subsection (a)(1); except that in no event shall an indemnity payment to any person exceed the fair market value of the pesticide owned by such person immediately before the notice referred to in subsection (a)(1).

"(2) SPECIAL RULE.—Notwithstanding any other provision of this Act, the Administrator may provide a reasonable time for use or other disposal of such pesticide. In determining the quantity of any pesticide for which indemnity shall be paid under this subsection, proper adjustment shall be made for any pesticide used or otherwise disposed of by such owner.

"SEC. 16. ADMINISTRATIVE PROCEDURE; JUDICIAL REVIEW.

"(a) DISTRICT COURT REVIEW.—Except as is otherwise provided in this Act, Agency refusals to cancel or suspend registrations or change classifications not following a hearing and other final Agency actions not committed to Agency discretion by law are judicially reviewable in the district courts.

"(b) REVIEW BY COURT OF APPEALS.—In the case of actual controversy as to the validity of any order issued by the Administrator following a public hearing, any person who will be adversely affected by such order and who had been a party to the proceedings may obtain judicial review by filing in the United States court of appeals for the circuit wherein such person resides or has a place of business, within 60 days after the entry of such order, a petition praying that the order be set aside in whole or in part. A copy of the petition shall be forthwith transmitted by the clerk of the court to the Administrator or any officer designated by him for that purpose, and thereupon the Administrator shall file in the court the record of the proceedings on which he based his order, as provided in section 2112 of title 28, United States Code. Upon the filing of such petition the court shall have exclusive jurisdiction to affirm or set aside the order complained of in whole or in part. The court shall consider all evidence of record. The order of the Administrator shall be sustained if it is supported by substantial evidence when considered on the record as a whole. The judgment of the court affirming or setting aside, in whole or in part, any order under this section shall be final, subject to review by the Supreme Court of the United States upon certiorari or certification as provided in section 1254 of title 28 of the United States Code. The commencement of proceedings under this section shall not, unless specifically ordered by the court to the contrary, operate as a stay of an order. The court shall advance on the docket and expedite the disposition of all cases filed therein pursuant to this section.

"(c) JURISDICTION OF DISTRICT COURTS.—The district courts of the United States are vested with jurisdiction specifically to enforce, and to prevent and restrain violations of, this Act.

"(d) NOTICE OF JUDGMENTS.—The Administrator shall, by publication in such manner as he may prescribe, give notice of all judgments entered in actions instituted under the authority of this Act.

72 Stat. 941;
80 Stat. 1323.

62 Stat. 928.

239

"SEC. 17. IMPORTS AND EXPORTS.

"(a) PESTICIDES AND DEVICES INTENDED FOR EXPORT.—Notwithstanding any other provision of this Act, no pesticide or device shall be deemed in violation of this Act when intended solely for export to any foreign country and prepared or packed according to the specifications or directions of the foreign purchaser, except that producers of such pesticides and devices shall be subject to section 8 of this Act.

"(b) CANCELLATION NOTICES FURNISHED TO FOREIGN GOVERNMENTS.—Whenever a registration, or a cancellation or suspension of the registration of a pesticide becomes effective, or ceases to be effective, the Administrator shall transmit through the State Department notification thereof to the governments of other countries and to appropriate international agencies.

"(c) IMPORTATION OF PESTICIDES AND DEVICES.—The Secretary of the Treasury shall notify the Administrator of the arrival of pesticides and devices and shall deliver to the Administrator, upon his request, samples of pesticides or devices which are being imported into the United States, giving notice to the owner or consignee, who may appear before the Administrator and have the right to introduce testimony. If it appears from the examination of a sample that it is adulterated, or misbranded or otherwise violates the provisions set forth in this Act, or is otherwise injurious to health or the environment, the pesticide or device may be refused admission, and the Secretary of the Treasury shall refuse delivery to the consignee and shall cause the destruction of any pesticide or device refused delivery which shall not be exported by the consignee within 90 days from the date of notice of such refusal under such regulations as the Secretary of the Treasury may prescribe: *Provided,* That the Secretary of the Treasury may deliver to the consignee such pesticide or device pending examination and decision in the matter on execution of bond for the amount of the full invoice value of such pesticide or device, together with the duty thereon, and on refusal to return such pesticide or device for any cause to the custody of the Secretary of the Treasury, when demanded, for the purpose of excluding them from the country, or for any other purpose, said consignee shall forfeit the full amount of said bond: *And provided further,* That all charges for storage, cartage, and labor on pesticides or devices which are refused admission or delivery shall be paid by the owner or consignee, and in default of such payment shall constitute a lien against any future importation made by such owner or consignee.

"(d) COOPERATION IN INTERNATIONAL EFFORTS.—The Administrator shall, in cooperation with the Department of State and any other appropriate Federal agency, participate and cooperate in any international efforts to develop improved pesticide research and regulations.

"(e) REGULATIONS.—The Secretary of the Treasury, in consultation with the Administrator, shall prescribe regulations for the enforcement of subsection (c) of this section.

"SEC. 18. EXEMPTION OF FEDERAL AGENCIES.

"The Administrator may, at his discretion, exempt any Federal or State agency from any provision of this Act if he determines that emergency conditions exist which require such exemption.

"SEC. 19. DISPOSAL AND TRANSPORTATION.

"(a) PROCEDURES.—The Administrator shall, after consultation with other interested Federal agencies, establish procedures and regula- Regulations.

240

tions for the disposal or storage of packages and containers of pesticides and for disposal or storage of excess amounts of such pesticides, and accept at convenient locations for safe disposal a pesticide the registration of which is canceled under section 6(c) if requested by the owner of the pesticide.

"(b) ADVICE TO SECRETARY OF TRANSPORTATION.—The Administrator shall provide advice and assistance to the Secretary of Transportation with respect to his functions relating to the transportation of hazardous materials under the Department of Transportation Act (49 U.S.C. 1657), the Transportation of Explosives Act (18 U.S.C. 831–835), the Federal Aviation Act of 1958 (49 U.S.C. 1421–1430, 1472 H), and the Hazardous Cargo Act (46 U.S.C. 170, 375, 416).

80 Stat. 944.
74 Stat. 808;
79 Stat. 286.
72 Stat. 775;
85 Stat. 481.
Contract
authority.

"SEC. 20. RESEARCH AND MONITORING.

"(a) RESEARCH.—The Administrator shall undertake research, including research by grant or contract with other Federal agencies, universities, or others as may be necessary to carry out the purposes of this Act, and he shall give priority to research to develop biologically integrated alternatives for pest control. The Administrator shall also take care to insure that such research does not duplicate research being undertaken by any other Federal agency.

"(b) NATIONAL MONITORING PLAN.—The Administrator shall formulate and periodically revise, in cooperation with other Federal, State, or local agencies, a national plan for monitoring pesticides.

"(c) MONITORING.—The Administrator shall undertake such monitoring activities, including but not limited to monitoring in air, soil, water, man, plants, and animals, as may be necessary for the implementation of this Act and of the national pesticide monitoring plan. Such activities shall be carried out in cooperation with other Federal, State, and local agencies.

"SEC. 21. SOLICITATION OF COMMENTS; NOTICE OF PUBLIC HEARINGS.

"(a) The Administrator, before publishing regulations under this Act, shall solicit the views of the Secretary of Agriculture.

"(b) In addition to any other authority relating to public hearings and solicitation of views, in connection with the suspension or cancellation of a pesticide registration or any other actions authorized under this Act, the Administrator may, at his discretion, solicit the views of all interested persons, either orally or in writing, and seek such advice from scientists, farmers, farm organizations, and other qualified persons as he deems proper.

Publication
in Federal
Register.

"(c) In connection with all public hearings under this Act the Administrator shall publish timely notice of such hearings in the Federal Register.

"SEC. 22. DELEGATION AND COOPERATION.

"(a) DELEGATION.—All authority vested in the Administrator by virtue of the provisions of this Act may with like force and effect be executed by such employees of the Environmental Protection Agency as the Administrator may designate for the purpose.

"(b) COOPERATION.—The Administrator shall cooperate with the Department of Agriculture, any other Federal agency, and any appropriate agency of any State or any political subdivision thereof, in carrying out the provisions of this Act, and in securing uniformity of regulations.

"SEC. 23. STATE COOPERATION, AID, AND TRAINING.

"(a) COOPERATIVE AGREEMENTS.—The Administrator is authorized to enter into cooperative agreements with States—

"(1) to delegate to any State the authority to cooperate in the enforcement of the Act through the use of its personnel or facilities, to train personnel of the State to cooperate in the enforcement of this Act, and to assist States in implementing cooperative enforcement programs through grants-in-aid; and

"(2) to assist State agencies in developing and administering State programs for training and certification of applicators consistent with the standards which he prescribes.

"(b) CONTRACTS FOR TRAINING.—In addition, the Administrator is authorized to enter into contracts with Federal or State agencies for the purpose of encouraging the training of certified applicators.

"(c) The Administrator may, in cooperation with the Secretary of Agriculture, utilize the services of the Cooperative State Extension Services for informing farmers of accepted uses and other regulations made pursuant to this Act.

"**SEC. 24. AUTHORITY OF STATES.**

"(a) A State may regulate the sale or use of any pesticide or device in the State, but only if and to the extent the regulation does not permit any sale or use prohibited by this Act;

"(b) such State shall not impose or continue in effect any requirements for labeling and packaging in addition to or different from those required pursuant to this Act; and

"(c) a State may provide registration for pesticides formulated for distribution and use within that State to meet special local needs if that State is certified by the Administrator as capable of exercising adequate controls to assure that such registration will be in accord with the purposes of this Act and if registration for such use has not previously been denied, disapproved, or canceled by the Administrator. Such registration shall be deemed registration under section 3 for all purposes of this Act, but shall authorize distribution and use only within such State and shall not be effective for more than 90 days if disapproved by the Administrator within that period.

"**SEC. 25. AUTHORITY OF ADMINISTRATOR.**

"(a) REGULATIONS.—The Administrator is authorized to prescribe regulations to carry out the provisions of this Act. Such regulations shall take into account the difference in concept and usage between various classes of pesticides.

"(b) EXEMPTION OF PESTICIDES.—The Administrator may exempt from the requirements of this Act by regulation any pesticide which he determines either (1) to be adequately regulated by another Federal agency, or (2) to be of a character which is unnecessary to be subject to this Act in order to carry out the purposes of this Act.

"(c) OTHER AUTHORITY.—The Administrator, after notice and opportunity for hearing, is authorized—

"(1) to declare a pest any form of plant or animal life (other than man and other than bacteria, virus, and other micro-organisms on or in living man or other living animals) which is injurious to health or the environment:

"(2) to determine any pesticide which contains any substance or substances in quantities highly toxic to man;

"(3) to establish standards (which shall be consistent with those established under the authority of the Poison Prevention Packaging Act (Public Law 91–601)) with respect to the package, container, or wrapping in which a pesticide or device is enclosed for use or consumption, in order to protect children and adults from serious injury or illness resulting from accidental ingestion 84 Stat. 1670. 15 USC 1471 note.

242

or contact with pesticides or devices regulated by this Act as well as to accomplish the other purposes of this Act;

"(4) to specify those classes of devices which shall be subject to any provision of paragraph 2(q) (1) or section 7 of this Act upon his determination that application of such provision is necessary to effectuate the purposes of this Act;

"(5) to prescribe regulations requiring any pesticide to be colored or discolored if he determines that such requirement is feasible and is necessary for the protection of health and the environment; and

"(6) to determine and establish suitable names to be used in the ingredient statement.

"SEC. 26. SEVERABILITY.

"If any provision of this Act or the application thereof to any person or circumstance is held invalid, the invalidity shall not affect other provisions or applications of this Act which can be given effect without regard to the invalid provision or application, and to this end the provisions of this Act are severable.

"SEC. 27. AUTHORIZATION FOR APPROPRIATIONS.

"There is authorized to be appropriated such sums as may be necessary to carry out the provisions of this Act for each of the fiscal years ending June 30, 1973, June 30, 1974, and June 30, 1975. The amounts authorized to be appropriated for any fiscal year ending after June 30, 1975, shall be the sums hereafter provided by law."

AMENDMENTS TO OTHER ACTS

SEC. 3. The following Acts are amended by striking out the terms "economic poisons" and "an economic poison" wherever they appear and inserting in lieu thereof "pesticides" and "a pesticide" respectively:

74 Stat. 1305.

(1) The Federal Hazardous Substances Act, as amended (15 U.S.C. 1261 et seq.) ;

84 Stat. 1670.

(2) The Poison Prevention Packaging Act, as amended (15 U.S.C. 1471 et seq.) ; and

52 Stat. 1040.

(3) The Federal Food, Drug, and Cosmetic Act, as amended (21 U.S.C. 301 et seq.).

EFFECTIVE DATES OF PROVISIONS OF ACT

SEC. 4. (a) Except as otherwise provided in the Federal Insecticide, Fungicide, and Rodenticide Act, as amended by this Act, and as otherwise provided by this section, the amendments made by this Act shall take effect at the close of the date of the enactment of this Act, provided if regulations are necessary for the implementation of any provision that becomes effective on the date of enactment, such regulations shall be promulgated and shall become effective within 90 days from the date of enactment of this Act.

Savings provision. 61 Stat. 163. 7 USC 135 note.

(b) The provisions of the Federal Insecticide, Fungicide, and Rodenticide Act and the regulations thereunder as such existed prior to the enactment of this Act shall remain in effect until superseded by the amendments made by this Act and regulations thereunder: *Provided,* That all provisions made by these amendments and all regulations thereunder shall be effective within four years after the enactment of this Act.

(c) (1) Two years after the enactment of this Act the Administrator shall have promulgated regulations providing for the registration and classification of pesticides under the provisions of this Act and thereafter shall register all new applications under such provisions.

(2) After two years but within four years after the enactment of this Act the Administrator shall register and reclassify pesticides registered under the provisions of the Federal Insecticide, Fungicide, and Rodenticide Act prior to the effective date of the regulations promulgated under subsection (c)(1).

61 Stat. 163.
7 USC 135
note.

(3) Any requirements that a pesticide be registered for use only by a certified applicator shall not be effective until four years from the date of enactment of this Act.

(4) A period of four years from date of enactment shall be provided for certification of applicators.

(A) One year after the enactment of this Act the Administrator shall have prescribed the standards for the certification of applicators.

(B) Within three years after the enactment of this Act each State desiring to certify applicators shall submit a State plan to the Administrator for the purpose provided by section 4(b).

(C) As promptly as possible but in no event more than one year after submission of a State plan, the Administrator shall approve the State plan or disapprove it and indicate the reasons for disapproval. Consideration of plans resubmitted by States shall be expedited.

(5) One year after the enactment of this Act the Administrator shall have promulgated and shall make effective regulations relating to the registration of establishments, permits for experimental use, and the keeping of books and records under the provisions of this Act.

(d) No person shall be subject to any criminal or civil penalty imposed by the Federal Insecticide, Fungicide, and Rodenticide Act, as amended by this Act, for any act (or failure to act) occurring before the expiration of 60 days after the Administrator has published effective regulations in the Federal Register and taken such other action as may be necessary to permit compliance with the provisions under which the penalty is to be imposed.

(e) For purposes of determining any criminal or civil penalty or liability to any third person in respect of any act or omission occurring before the expiration of the periods referred to in this section, the Federal Insecticide, Fungicide, and Rodenticide Act shall be treated as continuing in effect as if this Act had not been enacted.

Approved October 21, 1972.

LEGISLATIVE HISTORY:

HOUSE REPORTS: No. 92-511 (Comm. on Agriculture) and No. 92-1540
          (Comm. of Conference).
SENATE REPORTS:No. 92-838 (Comm. on Agriculture and Forestry) and
          NO. 92-970 (Comm. on Commerce).
CONGRESSIONAL RECORD:
    Vol. 117 (1971): Nov. 8,9, considered and passed House.
    Vol. 118 (1972): Sept.26, considered and passed Senate, amended.
                      Oct. 5, Senate agreed to conference report.
                      Oct. 12, House agreed to conference report.
WEEKLY COMPILATION OF PRESIDENTIAL DOCUMENTS:
    Vol. 8, No. 44 (1972): Oct. 21, Presidential statement.

Summary:

The Federal Environmental Pesticide Control Act of 1972 was enacted to amend the Federal Insecticide, Fungicide, and Rodenticide Act for the purpose of controlling the registration, manufacture, sale, and use of pesticides. The administrator of this act is authorized to prescribe regulations to carry out the provisions of this act. These regulations shall take into account the differences in concept and usage between various classes of pesticides. Pesticides regulated by another Federal agency, or are of the type unnecessary to be subjected to this Act, are exempt from the regulations of this Act.

Review Questions:

1. What are three specific purposes of the Federal Environmental Pesticide Control Act of 1972?

2. Identify one economic factor that aided the passage of the Federal Environmental Pesticide Control Act of 1972?

3. Identify one political factor that aided the passage of the Federal Environmental Pesticide Control Act of 1972?

4. Identify one social factor that aided the passage of the Federal Environmental Pesticide Control Act of 1972?

5. What effect has the Federal Environmental Pesticide Control Act of 1972 had on the chemical manufacturing industry? the agricultural industry?

6. What are two strengths of the Federal Environmental Pesticide Control Act of 1972 with regard to current pollution control problems?

7. What are two weaknesses of the Federal Environmental Pesticide Control Act of 1972 with regard to current pollution control problems?

radiation

Throughout his history, man has been exposed to cosmic and other naturally-occurring radiation. This natural background radiation still constitutes about 55 percent of the total radiation dose reaching the average American each year. The remainder comes from a variety of man-made radiation sources, including x-rays, the operating of nuclear power plants, and electronic devices in the home and workplace.

The potential benefits of successful application of nuclear and electro-magnetic technology are tremendous. However, since radiation can cause cancer or other injury to the body, and since the degree of risk is assumed to vary in direct proportion to the level of exposure, society has a responsibility to keep man-made exposures as low as possible.

Radiation generally classed as "environmental" is only a part of the problem. Levels far higher than are present in the environment today reach increasing numbers of people from "non-environmetal" sources. Medical uses of radiation, for example, now represent about 94 percent of all exposure to man-made radiation, or roughly 40 percent of all radiation sources to which the average person is exposed. Moreover, the last few years have brought increasing application of radiation in research and industrial processing as well as a phenomenal growth in the use of radiation-generating electronic products in the home and workplace.

As a source of environmental radiation, the use of nuclear energy to generate electric power has become an increasing focus of concern. Some 20 nuclear power plants are now in operation in the United States, and about 450 are expected to be in use by the year 1990. Small amounts of radiation are released into the environment from these reactors and from fuel reprocessing plants. These emissions constitute only about .003 percent of all man-made radiation to which even those persons living near the plants are exposed. However,

TOWARD A NEW ENVIRONMENTAL ETHIC, U.S. Government Printing Office: 1971-O — 443-062, pp. 20-21.

since any increase in radiation is believed to be accompanied by increased risk, even low-level exposures cannot be ignored.

The possibility of accidental release of large amounts of radioactivity from nuclear reactors also cannot be dismissed, even though safety has been stressed since the very inception of the nuclear power industry to make this possibility extremely remote.

The disposal of radioactive wastes from nuclear power generation and fuel reprocessing is a problem which may be expected to increase with the growth of the nuclear industry. At present, such wastes are buried or stored at carefully selected sites and a close watch is maintained to assure that leakage does not occur. About two million cubic feet of solid wastes of low-level radioactivity are interred in authorized burial grounds. Some 80 million gallons of stored high-level liquid wastes are in the process of being solidified through newly developed technology which will reduce their volume to one-tenth of the liquid form.

Fallout from weapons-testing prior to the 1963 atmospheric nuclear test ban treaty currently contributes about 3 percent of the man-made radiation to which Americans are exposed.

Health effects which may result from exposure to relatively large doses of ionizing radiation are well known: leukemia and other types of cancer, reduction in fertility, cataracts and other eye damage, acceleration of the aging process, and damage to reproductive cells. There is little understanding, however, of the long-term effects from repeated exposure to all forms of radiation at low-levels. A major potential hazard is damage to, or alteration of, human genes, since natural background radiation is believed to be one of the causes of natural mutation. It must be recognized, therefore, that the protracted release of even very low levels of long-lasting radioactivity from an increasing number of man-made sources has implications for human health which science has barely begun to explore.

The hazards associated with radiation, unlike those of other environmental pollutants, were dramatically illustrated long before widespread commercial application of radiation-producing technology took place. Strict governmental controls were imposed early, therefore, and the formal procedures and scientific bases for establishing and enforcing standards for protection against ionizing radiation have been the most comprehensive of any applied to environmental stresses. Even so, recent federal actions have been aimed at making doubly sure that the utmost precautions are observed:

• Under the reorganization plan establishing the United States Environmental Protection Agency, EPA assumed federal authority to set generally applicable environmental radiation standards. The Atomic Energy Commision (AEC) retains authority to implement and enforce EPA standards in the regulation of radioactive materials and nuclear facilities.

• When EPA was established a comprehensive review of existing radiation standards was underway to determine their adequacy. EPA, in cooperation with the U.S. Department of Health, Education, and Welfare, the Atomic Energy Commission and other Federal agencies, has continued the review, and it is scheduled for completion in 1972.

• Radioactive emissions from nuclear power reactors have constituted only a small fraction of the limits permitted under the radiation control standards which, as mentioned, are currently being reviewed. In 1971, in order to hold such emissions to the "lowest practicable levels," the AEC proposed new design and operating guidelines aimed at limiting emissions to 1/100th of the levels permitted under the standards.

• Under the provisions of the National Environmental Policy Act, EPA reviews all proposals of the Atomic Energy Commission which involve the siting, construction, and operation of nuclear facilities.

• EPA conducts research on the health impact of radiation from all sources, and monitors radiation in the environment.

Federal authorities have developed and have in partial operation an improved state-federal-industry system for monitoring environmental radiation sources to provide improved surveillance capability as the nuclear power industry expands.

248

ENERGY AND POLLUTION CONTROL

Pollution control legislation enacted at the Federal and State levels of government has had a direct effect upon the use of energy in the United States. The increased demand for gasoline, restrictions on the use of coal and high-sulfur residual oil, problems of refinery plant sitings, delays in the construction of the Alaska pipeline, and restrictions on off-shore oil drilling operations are but a few of the problems related to pollution control legislation.

This section, prepared by the American Petroleum Institute, covers thirteen points regarding the energy problem. These thirteen points are:

I.   Motor Gasoline

II.   Distillate Oil

III.   Residual Fuel Oil

IV.   U. S. Refinery Operations

V.   Crude Oil

VI.   Imports

VII.   U. S. Petroleum Exports

VIII. Natural Gas

IX.   Environmental Delays and Restrictions

X.   Industry Profits

XI.   Prices of Crude Oil and Gasoline

XII.   Free World Petroleum

XIII. Other Energy Sources

POINT SHEETS ON THE ENERGY PROBLEM, American Petroleum Institute, July 25, 1973.

I. MOTOR GASOLINE[1]

1. <u>Gasoline demand</u> in the United States has increased sharply over the past several years:

        1964 -- 4.4 million barrels daily[2]
        1969 -- 5.5 million barrels daily
        1970 -- 5.8 million barrels daily
        1971 -- 6.0 million barrels daily
        1972 -- 6.4 million barrels daily

The rise in consumption of gasoline in 1972 represented an increase of 6.0 per cent over 1971. Data for the first quarter of 1973 show an increase of 7.7 per cent in demand for motor gasoline over the corresponding period in 1972.

2. <u>Major causes for the rise in gasoline demand are:</u>

- Pollution controls on new cars requiring more gasoline (estimated at 300,000 barrels daily by the Office of Emergency Preparedness);

- Increased public purchase of gasoline-consuming power options on new cars (such as air conditioning -- 69 per cent, and automatic transmissions -- 90 per cent);

- Continued demand for large cars (although purchases of a greater proportion of smaller cars in recent months seem to indicate a shift in consumer buying);

- Record-breaking sales of new cars -- all of which have the new emission controls, and most of which have air conditioning and other power options (automobile production in June 1973 set an all-time record; automobile production in the first six months of 1973 totaled 6.18 million units, compared with 5.39 million units during the same period in 1972); and

- Increased away-from-home vacations (with resultant greater use of travel trailers, pleasure boats, snowmobiles, and other gasoline-consuming vehicles and equipment).

---

[1] Motor gasoline is that gasoline used in on-road and off-road vehicles and equipment, but excluding aviation gasoline.

[2] There are 42 U.S. gallons to a barrel. To convert to barrels per year, multiply by 365; to gallons per year: barrels x 42 x 365.

3. Motor gasoline production in the first six months
of 1973 has been higher than for the first six months of
any previous year. Production during the week ending June
1, 1973 set an all-time record of 49.351 million barrels.
And total production for the first six months of 1973 was
some 83 million barrels greater than for the comparable
period in 1972.

4. Motor gasoline stocks[1] are subject to seasonal
variations, caused by drawdowns (removals) to supplement
current production during high-use periods, and refinery
turnaround (when production is stopped to permit periodic
repairs to be made). In any consideration of fuel stocks,
it is important to relate those stocks with demand since
a "large" stock could be rapidly depleted in a period of
high-fuel consumption.

A comparison of the ratio of stocks on hand to domestic demand as of the last week in June 1973 and of the
equivalent week in each other year 1969 through 1972
shows[2]:

           1969 -- 192.6 million barrels or a 33-day supply
           1970 -- 210.9 million barrels or a 35-day supply
           1971 -- 211.4 million barrels or a 34-day supply
           1972 -- 207.6 million barrels or a 31-day supply
           1973 -- 205.2 million barrels or a 29-day supply

---

[1] The term "stocks" refers to inventories of fuel located
at refineries, bulk terminals, and in pipelines. These
inventories include stocks unavailable for shipment,
those available for current shipment, and those held in
reserve for shipment during the high-consuming season
as a supplement to then current production. Unavailable
stocks are those required for processing, held in tank
bottoms, in pipeline fill, and other equipment in order
to assure continuous operations.

[2] 1969-1973 data based on stock figures (from API) as of
the last week of June 1973 and the equivalent week of
each other year, and demand figures (from U.S. Bureau
of Mines) for the third quarter of each year (1969-
1972 actual, 1973 estimated).

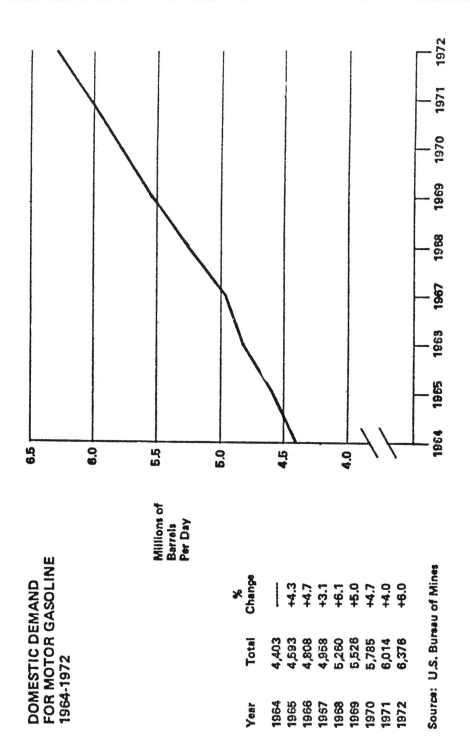

DOMESTIC DEMAND
FOR MOTOR GASOLINE
1964-1972

Millions of
Barrels
Per Day

Year	Total	% Change
1964	4,403	—
1965	4,593	+4.3
1966	4,808	+4.7
1967	4,958	+3.1
1968	5,260	+6.1
1969	5,528	+5.0
1970	5,785	+4.7
1971	6,014	+4.0
1972	6,376	+6.0

Source: U.S. Bureau of Mines

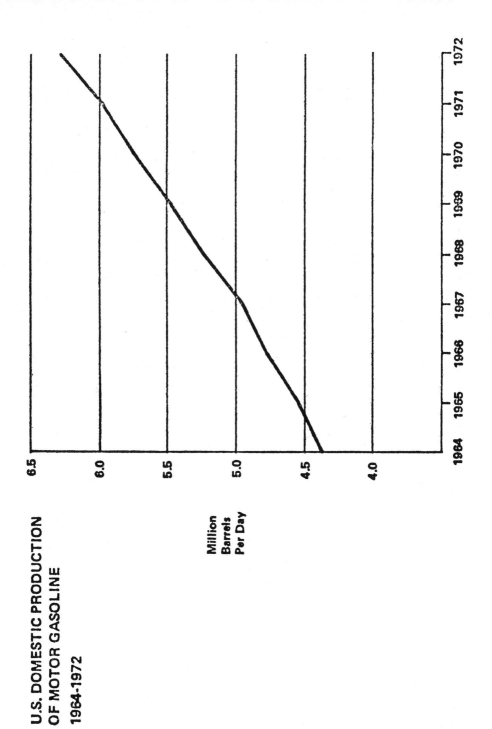

U.S. DOMESTIC PRODUCTION
OF MOTOR GASOLINE
1964-1972

Million
Barrels
Per Day

253

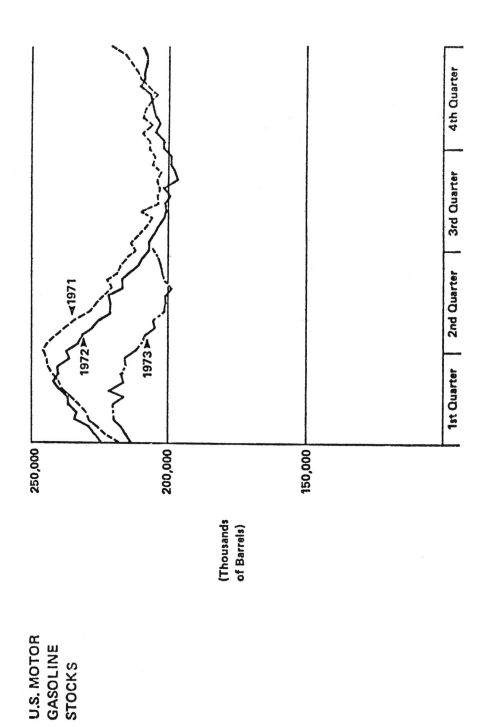

U.S. MOTOR
GASOLINE
STOCKS

(Thousands
of Barrels)

250,000

200,000

150,000

1st Quarter | 2nd Quarter | 3rd Quarter | 4th Quarter

◄1971

1972►

1973►

254

II. DISTILLATE OIL (HOME HEATING OIL AND DIESEL FUEL)

1. Distillate oil (primarily home heating oil and diesel fuel) demand has risen sharply in recent years:

        1964 -- 2.05 million barrels daily
        1969 -- 2.46 million barrels daily
        1970 -- 2.54 million barrels daily
        1971 -- 2.66 million barrels daily
        1972 -- 2.91 million barrels daily

The rise in consumption of distillate oil in 1972 represented an increase of 9.5 per cent over 1971. Data for the first quarter of 1973 show a continued increase of almost 3 per cent more in demand for distillate oil over the corresponding period in 1972.

2. A major cause for the rise in distillate oil demand was the increase in use of distillate by many electric utilities to generate electricity. This resulted from restrictions on the use of coal and high-sulfur residual oil in many areas for environmental reasons; the failure of programmed nuclear power plants to come on stream as anticipated -- again, in many instances, because of environmental problems; and the shortfall in natural gas, preventing its use in peak shaving periods.

Electric utility use of distillates during 1967 through 1972 has risen as follows:

        1967 --   8,000 barrels daily
        1968 -- 23,000 barrels daily
        1969 -- 33,000 barrels daily
        1970 -- 68,000 barrels daily
        1971 -- 97,000 barrels daily
        1972 --186,000 barrels daily

The 1972 level of distillate use by electric utilities was almost double that of 1971, and nearly triple that of 1970. As a point of comparison, daily distillate use by electric utilities in the U.S. was equivalent to about:

- 80 per cent of the total daily distillate use of all of America's railroads in 1971; or

- 40 per cent of the daily diesel fuel use by all U.S. trucking in 1971; or

- 125 per cent of the daily use of diesel fuel by U.S. farm machinery in 1971.

Other major causes in the demand increase for distillate fuels in the 1972-1973 fall and winter included:

- The switch by many manufacturing and commercial

255

businesses to distillate fuels from natural
gas -- when supplies of this latter fuel were
curtailed or in danger of being curtailed --
and from other fuels for environmental
reasons; and

- The exceptionally wet fall of 1972, coupled
with an early cold spell, resulted in an early
and large demand for distillate, diesel and
propane fuels for crop drying in the Midwest.

3. Distillate oil production in the first six months
of 1973 has been higher than for the first six months of
any previous year. Production in the week ending February
2, 1973, set an all-time record of 22.06 million barrels.
And total production of distillate oil over the first six
months of 1973 was 501 million barrels -- more than 32
million barrels over the comparable period in 1972.

4. Distillate oil stocks,[1] like those of gasoline,
vary significantly with the season, and are affected by
increased demand for motor gasoline (causing a later start
on increased distillate production), fall-winter demand
exceeding current production levels, and the unavailabil-
ity of crude oil at the refinery. The variation in distillate stocks from year to year is evident from the
following table showing the ratio of stocks on hand to
domestic demand as of the last week of June 1973 and the
equivalent week of each other year 1969 through 1972[2]:

```
1969 -- 132.6 million barrels or a 48-day supply
1970 -- 135.3 million barrels or a 48-day supply
1971 -- 149.4 million barrels or a 51-day supply
1972 -- 129.3 million barrels or a 37-day supply
1973 -- 139.1 million barrels or a 39-day supply
```

---

[1] The term "stocks" refers to inventories of fuel located
at refineries, bulk terminals, and in pipelines. These
inventories include stocks unavailable for shipment,
those available for current shipment, and those held
in reserve for shipment during the high-consuming
season as a supplement to then current production.
Unavailable stocks are those required for processing,
held in tank bottoms, in pipeline fill, and other
equipment in order to assure continuous operations.

[2] 1969-1973 data based on stock figures (from API) as of
the last week of June 1973 and the equivalent week of
each other year, and demand figures (from U.S. Bureau
of Mines) for the fourth quarter of each year (1969-
1972 actual, 1973 estimated.)

Heavy demand in the fall of the year normally causes a drawdown of distillate oil stocks, substantially narrowing the margin between the combination of stocks and production, and demand.

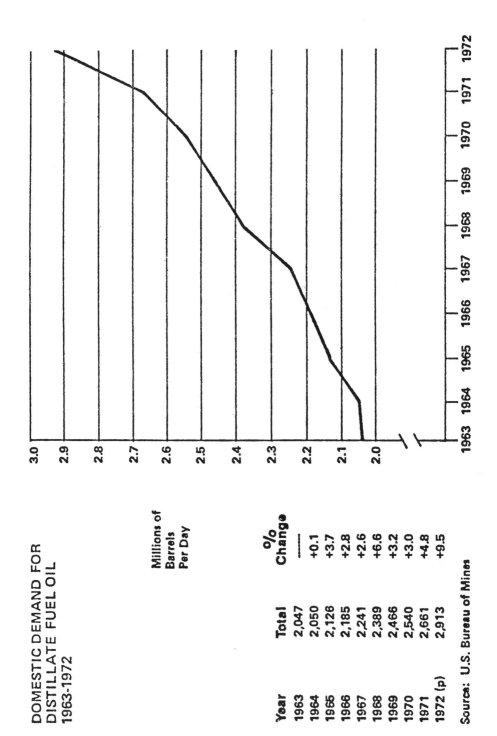

DOMESTIC DEMAND FOR
DISTILLATE FUEL OIL
1963-1972

Millions of
Barrels
Per Day

Year	Total	% Change
1963	2,047	—
1964	2,050	+0.1
1965	2,126	+3.7
1966	2,185	+2.8
1967	2,241	+2.6
1968	2,389	+6.6
1969	2,466	+3.2
1970	2,540	+3.0
1971	2,661	+4.8
1972 (p)	2,913	+9.5

Source: U.S. Bureau of Mines

SUPPLY OF DISTILLATE
FUEL OIL
1963-1972

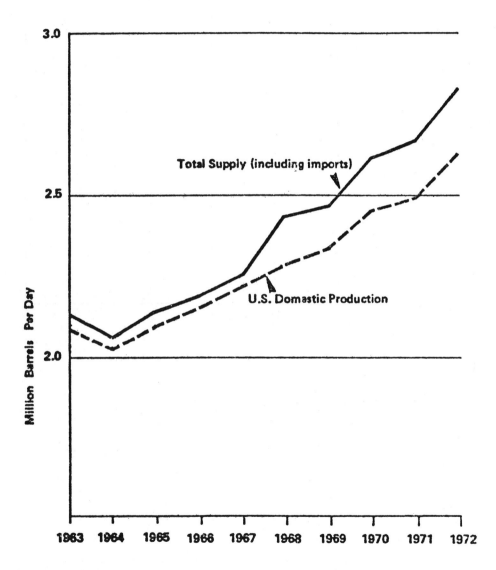

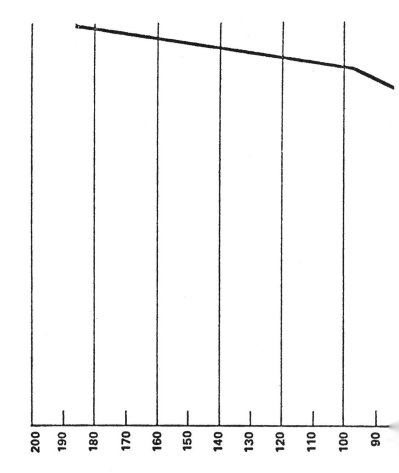

200 — 190 — 180 — 170 — 160 — 150 — 140 — 130 — 120 — 110 — 100 — 90

SALES OF DISTILLATE
FUEL OIL TO
ELECTRIC PUBLIC UTILITY
POWER PLANTS
1963-1972

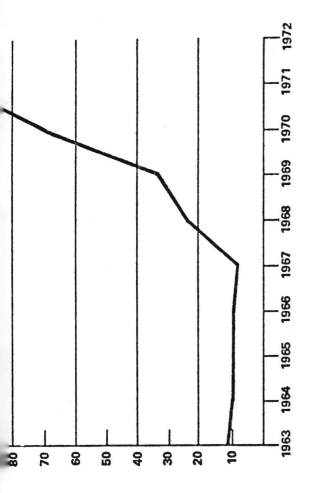

Source: U.S. Bureau of Mines

261

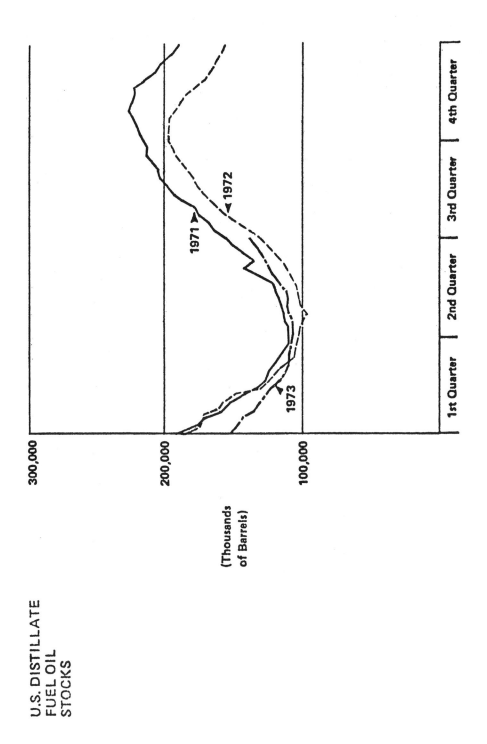

U.S. DISTILLATE
FUEL OIL
STOCKS

(Thousands
of Barrels)

300,000

200,000

100,000

1st Quarter | 2nd Quarter | 3rd Quarter | 4th Quarter

1971

1972

1973

III. RESIDUAL FUEL OIL

1. Demand for residual oil (primarily industrial heating oil and fuel for electric utilities) has risen sharply in recent years:

            1964 -- 1.52 million barrels daily
            1969 -- 1.98 million barrels daily
            1970 -- 2.20 million barrels daily
            1971 -- 2.30 million barrels daily
            1972 -- 2.53 million barrels daily

In 1972, demand for residual fuel increased 10.1 per cent from the 1971 level. Data for the first quarter of 1973 showed an 11.3 per cent increase from the corresponding period in 1972.

2. Because residual generally sold for less than the domestic crude oil from which it would be made, domestic production of residual oil decreased. In order to meet rising demand, imports spurted. In 1964, residual oil imports were 808,000 barrels daily and accounted for 53 per cent of domestic demand. By 1972, imports had more than doubled to reach 1.74 million barrels daily, and represented 69 per cent of demand.

3. Historically, a very high percentage of imported residual oil has come from Western Hemisphere sources (in 1972 about 90 per cent). However, an increasing amount of imported residual oil is refined in the Caribbean from African or Middle Eastern crude oil.

4. The East Coast of the United States is heavily dependent on residual oil imports. In 1972, the East Coast received more than 96 per cent of all residual oil imported into the United States. These imports provided almost 90 per cent of the East Coast's residual oil needs.

5. With natural gas supplies limited and environmental restrictions reducing coal availability, electric utilities are increasingly turning to oil to fill their fuel needs. In 1964, electric utilities used 267,000 barrels of residual oil daily. By 1972, electric utilities' demand for residual oil had increased almost fivefold to 1.22 million barrels daily.

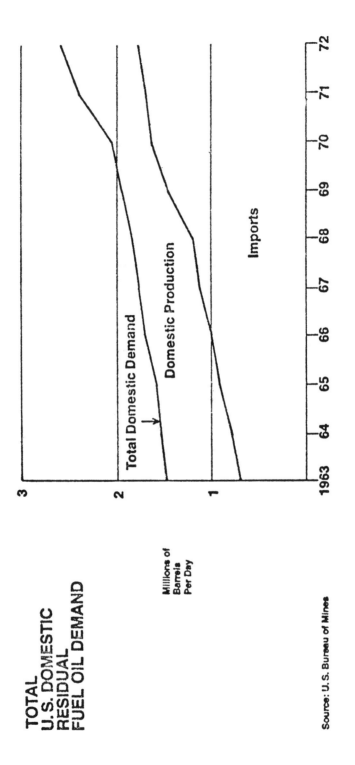

**TOTAL
U.S. DOMESTIC
RESIDUAL
FUEL OIL DEMAND**

Millions of
Barrels
Per Day

Total Domestic Demand

Domestic Production

Imports

3

2

1

1963 64 65 66 67 68 69 70 71 72

Source: U. S. Bureau of Mines

Demand		
Year	Total	% Change
1963	1,467	—
1964	1,515	+3.3
1965	1,608	+6.1
1966	1,716	+6.7
1967	1,786	+4.1
1968	1,826	+2.2
1969	1,978	+8.3
1970	2,204	+11.4
1971	2,296	+4.2
1972	2,529	+10.1

Imports		
Year	Total	% Change
1963	747	—
1964	808	+8.2
1965	946	+17.1
1966	1,032	+9.1
1967	1,085	+5.1
1968	1,120	+3.2
1969	1,265	+12.9
1970	1,528	+20.8
1971	1,583	+3.6
1972	1,742	+10.0

IV. U.S. REFINERY OPERATIONS

1. <u>There are about 250 refineries</u> (including asphalt producing plants, which operate seasonally) in the U.S., with a total operating capacity to refine of about 13.6 million barrels of crude oil per day. These refineries are operating at the highest level possible under current circumstances.

A survey taken the week ending March 30, 1973 of 174 refineries representing 95.2 per cent of the nation's total crude oil refining capacity showed that 78 were operating at or above their rated capacity and that 96 were operating below rated capacity. On the average, they operated at about 93 per cent of their rated capacity.

2. <u>Reasons reported by these refineries for not operating at capacity</u>[1] at the time of the survey included equipment shutdown for turnaround, maintenance and repair, and inability to obtain a type of crude oil suitable for the particular refinery.

- It should be noted that refineries normally shut down various operating units at some point during the spring of each year for a thorough inspection and necessary maintenance of the facility. Frequency of turnarounds varies with the individual refinery -- some need major maintenance after only a few months of around-the-clock operation, and others need turnaround as infrequently as once a year.

- Almost all refineries are designed to process a specific type of crude oil. For example, a refinery designed to process sweet (low-sulfur content) crude oil cannot process only sour (high-sulfur content) crude oil, without danger of seriously damaging the equipment.

3. <u>Expansion of refinery capacity</u> in the U.S. is essential to meet future petroleum demand. It is estimated that -- to meet demand by 1985 -- the equivalent of 60 new refineries, with a total daily capacity of 9 million barrels must be constructed. Some expansion of refining capacity at existing refineries is currently under way and more has been announced or is planned. However, not one new refinery is now under construction.

- In the 5-year period, 1968 through 1972, 1.9 million barrels a day were added to U.S.

---

[1] See the attached summary of the results of the refinery survey.

domestic refining capacity.

- During the same period, domestic demand for product increased by almost 3 million barrels daily.

4. Factors inhibiting the construction of new refining capacity in the U.S. include:

- Problems of siting -- for example, the State of Delaware recently passed legislation interdicting the construction of refineries along its coastline, and several other states are considering similar legislation.

- Questions of supply and type of crude oil -- it is vitally important that a company planning expansion or construction of refining capacity be assured that both the type and quantity of crude oil required will be available at the time the refinery is scheduled to become operable and for the practical life of the facility.

- Future product specifications -- just as refineries are designed to process certain crude oils, they are designed to produce certain petroleum products, within the limited range built in to provide for seasonal output variations. Therefore, before a refinery can be constructed, the company must know -- within reasonable ranges -- what the future government and automobile manufacturer specifications will be for the product. For example, what octane gasoline must be produced to meet government standards at the time the plant is to become operational, and for a reasonable period thereafter.

- The economic feasibility of construction -- problems of providing the enormous amounts of capital, when the cost of a new, 150,000 barrels a day refinery can exceed $200 million. Such expenditures can be made only if the costs are recoverable in the marketplace.

SUMMARY OF API SURVEY OF UTILIZATION OF OPERABLE REFINERY
CAPACITY IN THE U. S. DURING WEEK ENDED MARCH 30, 1973
(Barrels of 42 gallons)

Daily Average

Operable Refinery Capacity reported to API as of December 31, 1972	13,556,312
Operable refinery capacity reported by respondents to survey questionnaire	12,904,013
% of total operable capacity	95.2
Crude runs reported	12,036,558
Crude runs vs. operable capacity	-867,455
% utilization of operable capacity	93.3
Operable capacity of companies whose crude runs exceeded capacity	5,982,957
Crude runs reported	6,345,350
Crude runs in excess of operable capacity	+362,393
% utilization of operable capacity	106.1
Operable capacity of companies whose crude runs were below capacity	6,921,056
Crude runs reported	5,691,208
Crude runs below operable capacity	-1,229,848
% utilization of operable capacity	82.2

Reasons for running below capacity:

(a)	Shut down for turnaround, mechanical repairs, explosion and fire	531,710
(b)	Lack of crude oil	238,685
(c)	Lack of sweet crude oil	139,847
(d)	Crude unit metallurgy	0
(e)	Necessity of running grades of crude heavier than normally used	11,168
(f)	Processing oils other than crude	92,524
(g)	Downstream capacity limitations	58,610
(h)	Environmental constraints on refinery operations	15,000
(i)	Asphalt plant shutdown for seasonal reasons	57,747
(j)	Reported capacity overstated	15,000
(k)	Unavailability of other raw materials	1,491

(l) Other reasons:

(1)	Flood conditions	4,300
(2)	Crude receipt limited pending construction of new wharf now awaiting U.S. Govt. approval	41,279
(3)	Weather delayed tanker	6,800
(4)	Other misc. reasons	15,687

68,066

Total   1,229,848

V. CRUDE OIL

1. U.S. proved crude oil reserves[1], as of December 31, 1972 (including the Alaska North Slope) were estimated to be 36.3 billion barrels. This represents a decrease of 1.7 billion barrels from December 31, 1971.

- In addition to the 36.3 billion barrels of existing proved reserves, 99.9 billion barrels of crude oil have already been produced from U.S. petroleum reservoirs since the founding of the domestic petroleum industry in 1859.

2. The ratio of proved reserves to production has been in a general decline for more than a decade. Excluding the reserves on the Alaska North Slope[2], proved reserves were down to 26.7 billion barrels at the end of 1972, or about 8 times current production levels (as compared to 31.7 billion barrels in 1961, when reserves were 12½ times the then existing rate of production).

3. Demand for crude oil in the U.S. rose from 11.1 million barrels per day in 1971, to 11.7 million barrels per day in 1972. Domestic production accounted for 9.5 million barrels daily in 1972 (81 per cent of the demand), and imports accounted for 2.2 million barrels a day (19 per cent of the total demand for crude oil).

4. The major oil fields in the U.S. have been producing at their maximum efficient rate (MER). Wells in Louisiana have been operating at MER since August 1972, and those of Texas since April 1972. If wells were allowed to produce beyond MER, the natural pressures within the earth are too rapidly dissipated, with a consequent reduction in the amount of oil that can be recovered.

5. Crude oil stocks in the U.S. were at 254.4 million barrels on June 29, 1973. This compared with stocks of 274.5 million barrels on the same day in 1972, and 281.2 million barrels on the same day in 1971. Stocks of crude oil fluctuate throughout the year, and generally rise in mid-spring, while refineries are undergoing repair.

---

[1] Proved reserves are those quantities of oil underground which, with reasonable certainty, are recoverable from known reservoirs in the future, under existing economics and operating conditions.

[2] North Slope oil is excluded, since it cannot be brought to market until such time as a means of transport is in operation.

U.S. PROVED RESERVES OF CRUDE OIL END OF YEAR 1963-1972*

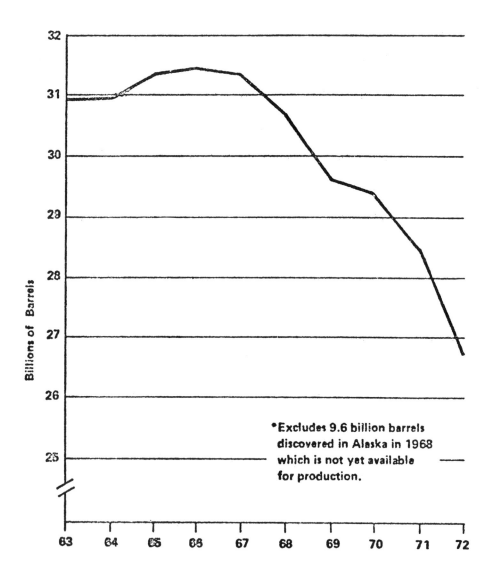

*Excludes 9.6 billion barrels discovered in Alaska in 1968 which is not yet available for production.

U.S. CRUDE
OIL STOCKS

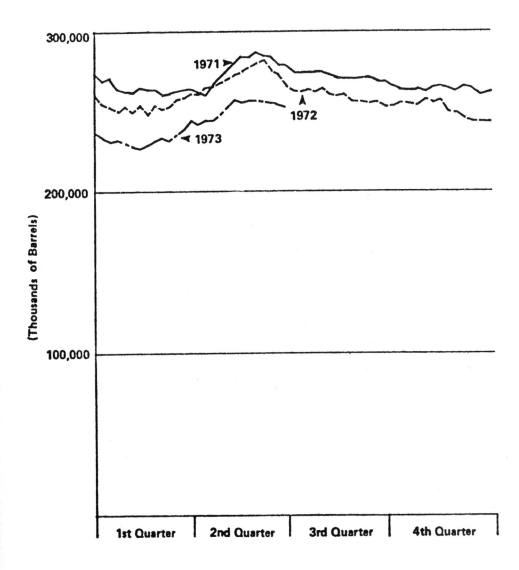

VI. IMPORTS

1. The U.S. is importing over one-third of the oil
and oil products it uses -- over 6 million barrels daily
during the first quarter of 1973. As a point of comparison, oil imports were in:

          1956 -- 1.4 million barrels per day.
          1971 -- 3.9 million barrels per day.
          1972 -- 4.7 million barrels per day.

2. The major sources of crude oil imports into the
U.S. during 1972 were:

          Canada   -- 854,000 barrels daily
          Venezuela   -- 255,000 barrels daily
          Nigeria   -- 242,000 barrels daily
          Saudi Arabia   -- 174,000 barrels daily
          Indonesia   -- 163,000 barrels daily
          Iran   -- 136,000 barrels daily
          Libya   -- 109,000 barrels daily
          Algeria   --   87,000 barrels daily
          Kuwait   --   36,000 barrels daily

3. Future demand for imports, reported by the National Petroleum Council, shows that -- if present explor-
ation and petroleum price trends continue, along with
present tax, regulatory and governmental policies toward
the petroleum industry -- by 1985, the U.S. would have to
depend on foreign sources for well over one-half of its
liquid petroleum supplies, and nearly 30 per cent of its
natural gas.

4. Changing trends in sources of crude oil and
product imports have become apparent. In 1971, Canada
and Venezuela supplied 47.9 per cent of our oil imports,
and the whole Western Hemisphere 78.4 per cent. In 1972,
the Western Hemisphere provided only 71.6 per cent of our
imports. Both Canada and Venezuela have about reached
the maximum amounts they can export to the U.S., because
of field limitations and national demand within those
countries. A major shift in imports to Eastern Hemisphere
sources took place in 1972. And, it is probable that,
without significant Western Hemisphere oil discoveries,
U.S. reliance on Middle East and Africa oil will substantially increase. Most Middle East crude oil has a high-
sulfur content.

5. Worldwide demand for crude oil during the first
quarter of 1973 continued to grow at a rapid pace, according to The Chase Manhattan Bank. Preliminary data for
the period indicate total oil used exceeded 53 million
barrels a day. In the U.S., consumption was 6 per cent
greater than for the corresponding period in 1972, while
demand in the rest of the world (excluding Russia, China
and Eastern Europe) grew almost twice as fast as in the U.S.

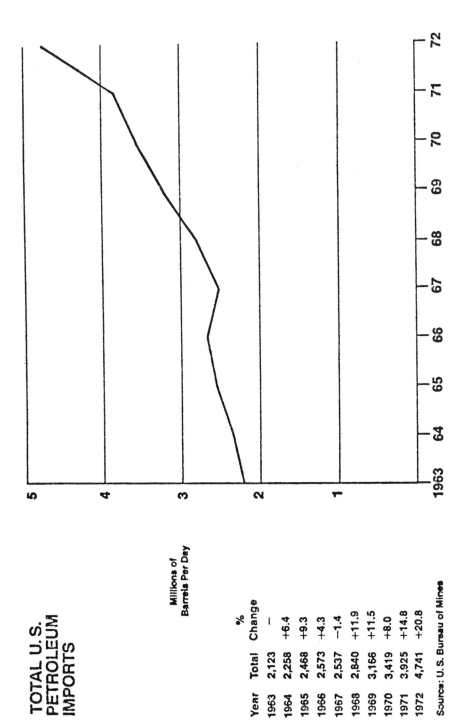

TOTAL U.S.
PETROLEUM
IMPORTS

Millions of
Barrels Per Day

Year	Total	% Change
1963	2,123	—
1964	2,258	+6.4
1965	2,468	+9.3
1966	2,573	+4.3
1967	2,537	–1.4
1968	2,840	+11.9
1969	3,166	+11.5
1970	3,419	+8.0
1971	3,925	+14.8
1972	4,741	+20.8

Source: U. S. Bureau of Mines

273

TOTAL U.S. PETROLEUM IMPORTS BY SOURCE 1963-1972

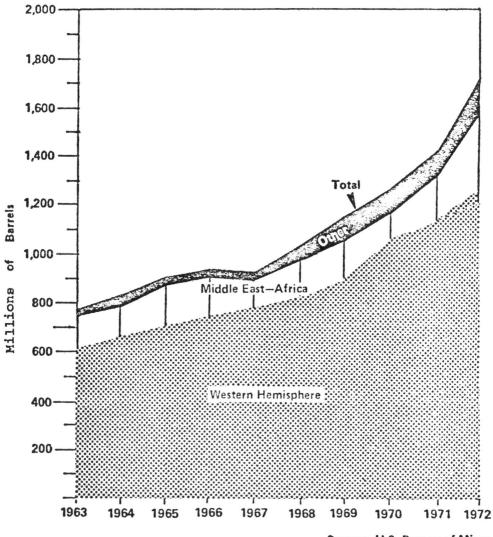

Source: U.S. Bureau of Mines

VII. U.S. PETROLEUM EXPORTS[1]

1. <u>U.S. petroleum exports in 1972</u>, according to U.S. government data, totaled 222,000 barrels a day. This total represents less than 1.4 per cent of the domestic petroleum product demand of almost 16.4 million barrels daily.

Most of the exports could not be economically used in domestic markets. Almost 86 per cent of the export total consisted of four product categories:

- <u>Coke</u> -- 85,000 barrels daily (38 per cent of petroleum exports) -- a surplus refining product sent primarily to Japan and Canada for use in metallurgical industries;

- <u>Lubricating oils</u> -- 41,000 barrels daily (18 per cent of petroleum exports) -- an export to a large extent unavailable outside of the U.S., and for which there is no prospect of a domestic shortage;

- <u>Residual oil</u> -- 33,000 barrels daily (15 per cent of petroleum exports) -- a surplus generated primarily on the West Coast, and for which there is no economical means by which it can be shipped to East Coast markets; and

- <u>Liquefied petroleum gases</u> -- 31,000 barrels daily (14 per cent of the petroleum exports) -- about half of which was transported by truck to Mexico, with no economical means of pipeline transportation to U.S. markets. A substantial increase in propane exports to Japan during the first four months of 1973 resulted from contracts entered into in 1972, when there was no domestic propane shortage.

2. <u>Distillate oil and motor gasoline exports</u> comprise only a minuscule portion of U.S. petroleum product exports. In 1972, distillate fuel oil exports totaled 3,300 barrels a day (1.5 per cent of exports, and 0.1 per cent of domestic demand for this product). These shipments went to diverse destinations, mainly on long-term contract.

---

[1] The following information is offered to refute allegations that the domestic petroleum supply problem is aggravated by the export of petroleum products.

Motor gasoline exports totaling 1,200 barrels daily
(0.5 per cent of exports, and 0.02 per cent of domestic
demand) went mainly to Mexico. While it is difficult to
trace these minor quantities, some was shipped to Baja
California -- where no other petroleum sources are available -- and some to northern Mexico.

VIII. NATURAL GAS

1. U.S. proved reserves of natural gas[1], as of
December 31, 1972 (including the Alaska North Slope) were
estimated to be 266.1 trillion cubic feet. This represents a decrease of 12.7 trillion cubic feet from December
31, 1971. At the end of 1972, natural gas reserves were
at about the same level as at year-end 1961 (266.3 trillion
cubic feet).

- In addition to the 266.1 trillion cubic feet
  of existing proved reserves, some 450 trillion
  cubic feet of natural gas have already been
  produced from U.S. petroleum reservoirs since
  1920.

2. The ratio of proved reserves to production has
declined steadily over the past 10 years. Excluding
reserves on Alaska's North Slope[2], proved reserves were
240.1 trillion cubic feet at the end of 1972, or about
11 times current production, as compared with 272.3
trillion cubic feet in 1962, when reserves were 20 times
annual production.

3. Demand for natural gas in the U.S. rose from
14.1 trillion cubic feet in 1962 to more than 23.5 trillion
cubic feet in 1972. Domestic production accounted for 22.9
trillion cubic feet (97 per cent of demand) and imports
accounted for somewhat over one trillion cubic feet (three
per cent of demand). To the extent that domestic supplies
of natural gas must be supplemented from foreign sources,
imports will increasingly consist of high-cost liquefied
natural gas.

4. The interdependence of energy sources was
illustrated last winter when natural gas shortages forced
curtailments of this fuel for some schools, commercial
establishments and other purchasers of gas on interruptible
contracts. (Interruptible contracts permit sales to be
suspended during peak periods of demand for users able to
switch to other fuels. Such contracts cover some 23 per
cent of total gas consumption, excluding field use.) These
curtailments diverted demand to oil.

---

[1] Proved reserves are those quantities of gas underground
which, with reasonable certainty, are recoverable from
known reservoirs in the future, under existing economic
and operating conditions.

[2] North Slope gas is excluded, since it cannot be brought
to market until such time as a means of transport is
available.

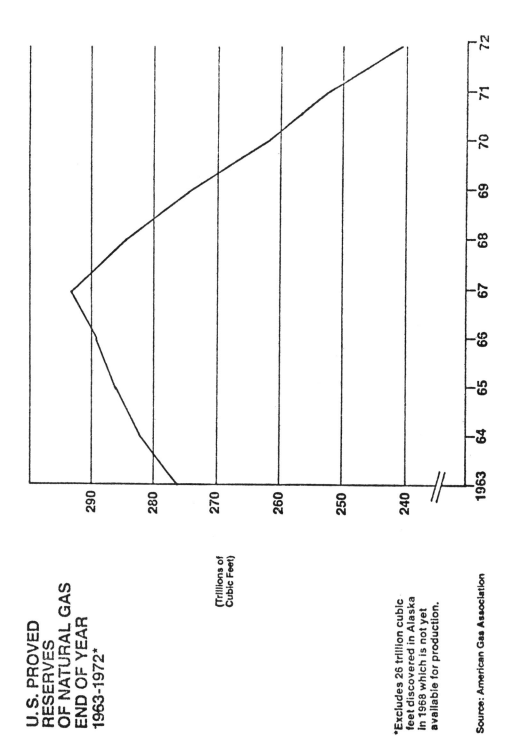

U.S. PROVED
RESERVES
OF NATURAL GAS
END OF YEAR
1963-1972*

(Trillions of
Cubic Feet)

*Excludes 26 trillion cubic
feet discovered in Alaska
in 1968 which is not yet
available for production.

Source: American Gas Association

IX. ENVIRONMENTAL DELAYS AND RESTRICTIONS

1. Environmental delays and restrictions have had their effect on virtually all phases of energy production in the U.S. These delays -- and in some cases, bans -- generated by environmental groups or by the government because of environmental considerations have made it increasingly difficult for the energy industries to meet the energy goals of the nation. While not making a judgment from the environmental point of view, it is clear that their impact on petroleum production has been substantial, and has served both to increase demand for oil products and to thwart development of critically needed new supplies.

2. Delays in constructing the Alaska pipeline have effectively prevented starting the delivery of the nearly 10 billion barrels of crude oil discovered at Prudhoe Bay, Alaska, in 1968 -- the largest petroleum discovery ever made in the Western Hemisphere. All alternatives for transporting the oil have been considered, including a test program for tanker transport through the Northwest Passage, and routing a pipeline down the Mackenzie River Valley, through Canada to the U.S. Midwest. These studies have conclusively shown that the most feasible method of transporting the oil is by pipeline. And both industry and government studies have concluded that the best route -- for environmental, time-frame, and economic reasons -- would be a trans-Alaska pipeline, with tanker shipment from the Southern Alaska port of Valdez.

An intensive environmental study was conducted by the federal government, and environmental protection stipulations on construction and operation of the line were drawn up. Yet the oil still remains underground at Prudhoe Bay. Even if a permit to construct the pipeline were granted today, it would be 1977 -- at the earliest -- before the oil could begin to be moved to market. The proposed pipeline has a design capacity of 2 million barrels daily throughput, although the actual throughput, based on known reserves, will probably be 1.5 million barrels daily.

3. Marine drilling operations provided about 17 per cent of our domestic crude oil and 15 per cent of our domestic natural gas in 1972. And it is estimated that the U.S. Outer Continental Shelf contains up to 190 billion barrels of crude oil and 1,100 trillion cubic feet of natural gas, recoverable under present technology and economics.

- Lease sales and schedules in the Gulf of
  Mexico were temporarily delayed because
  of environmental suits challenging the
  adequacy of the government environmental
  impact study. However, the environmental
  groups, which had blocked the earlier sale,
  did not challenge later sales.

- Sales of new leases in the Santa Barbara (California) Channel were suspended by the federal government, and a moratorium was placed on further drilling and production from 35 tracts previously sold in the channel. The moratorium continues today, although scientific studies of the effects of the 1969 oil spill there, which was the basis for the suspension of activities, found no evidence of permanent ecological damage from the spill, and demonstrated that the area has recovered well.

4. Opposition to refinery and petroleum facilities has been a contributing factor in construction delays. In at least one case, environmental opposition prevented the construction of a large new refinery, which was planned -- in spite of economics, supply, and product specification problems.

5. Automotive emission controls already on automobiles are consuming an additional 300,000 barrels of gasoline a day, and could rise to 2 million barrels daily by 1980, according to the Office of Emergency Preparedness.

6. Failure of the government to issue any construction or operating permits to nuclear power plants between early 1971 and mid-1972 -- for environmental and other reasons -- is estimated to have resulted in the use of the equivalent of hundreds of thousands of barrels of oil daily to generate power by alternate fuels.

7. Elimination of coal from some traditional markets, because of opposition to mining operations and to the burning of high-sulfur coal, has forced many users, including electric utilities, to switch to alternate fuels -- such as distillate oil.

X. INDUSTRY PROFITS

1. Average profits for the petroleum industry over the past 10 years (1963 through 1972) have been below the average for all-manufacturing. For this period, the average return on net worth for the petroleum industry was 11.8 per cent. The figure for all-manufacturing was 12.2 per cent. In only three of the past 10 years (1967, 1970, 1971) did petroleum profits rise above the all-manufacturing level.

2. Preliminary figures for 1972, from the First National City Bank of New York, indicate that petroleum profits averaged 10.8 per cent of net worth. The all-manufacturing figure was 12.2 per cent. Petroleum profits in 1972 showed a drop of 0.4 of a percentage point from the previous year, while the 1972 return for all-manufacturing increased by 1.3 percentage points.

3. Preliminary data for the first quarter of 1973, compiled by First National City Bank -- and based on a sample of 45 petroleum companies -- indicate that petroleum company profits were up 27.3 per cent from the first quarter of 1972. This increase reflects recovery from 1972's depressed first quarter profit levels, which were 3.1 per cent under the 1971 first quarter.

By comparison, average first quarter 1973 profits for the bank's sample of 1,045 manufacturing corporations were 31.0 per cent higher than for the first quarter of 1972.

XI. PRICES OF CRUDE OIL AND GASOLINE

1. Average annual crude oil prices, over the 10-year period, 1963 through 1972, rose from $2.89 per barrel to $3.40 per barrel. This increase of 17.6 per cent compares with a 26.0 per cent increase in the average annual whole-sale prices of all commodities, as reported by the U.S. Bureau of Labor Statistics for the same 10-year period. Thus, on a constant-dollar basis, the wholesale price of crude oil declined slightly during the decade.

The May 1973 average price of crude oil showed an increase of 25.3 per cent over the average price for 1963. Figured on the same basis, the wholesale prices of all commodities rose 41.3 per cent.

2. The stability of crude oil prices stands out in sharper relief when looked at in the perspective of rising costs facing the petroleum industry. For example, between 1963 and 1972, the U.S. Bureau of Labor Statistics reported that on an average annual basis:

- Oil well casing prices rose 44.6 per cent;

- Oil field machinery prices went up 36.4 per cent; and

- Average hourly wages in oil and gas production climbed 52.7 per cent.

3. The average annual wholesale price of gasoline, the industry's major product, in 1972, stood at 17.72 cents per gallon -- exclusive of taxes. This was 16.4 per cent above the ex tax wholesale price of gasoline in 1963, which averaged 15.22 cents per gallon. This in-crease, reflecting higher crude oil and refining costs, was also substantially below the rise in the overall wholesale price index.

In May 1973, the average wholesale price of regular grade gasoline, ex tax, was 19.21 cents per gallon reflecting an increase from the 1963 average of 26.2 per cent. Figured on the same basis, the wholesale prices of all commodities rose 41.3 per cent.

4. Average annual service station price of regular grade gasoline, ex tax, rose from 20.11 cents per gallon to 24.46 cents per gallon between 1963 and 1972, an increase of about 22 per cent, as compared to the 36.6 per cent increase in the prices of all consumer goods and services.

The May 1973 average of retail service station prices of regular grade gasoline, ex tax, rose 31.7 per cent over the average for 1963. Figured on the same basis, the retail prices of all consumer goods and services rose 43.4 per cent.

5.   The average annual wholesale price of distillate, according to the U.S. Bureau of Labor Statistics' Wholesale Price Index, rose 18.5 per cent from 1963 to 1972.   This compares with an increase of 26 per cent in the average annual wholesale price of all commodities.

The May 1973 average wholesale price of distillate showed an increase of 48.3 per cent over the average price for 1963.   Figured on the same basis, the wholesale prices of all commodities rose 41.3 per cent.

6.   The average annual retail price of No. 2 (home heating) oil,   according to the U.S. Bureau of Labor Statistics' Consumer Price Index, rose 23.3 per cent from 1963 to 1972, while the retail price of all consumer goods and services increased 36.6 per cent.

The May 1973 average retail price of No. 2 oil showed an increase of 34.6 per cent over the average price for 1963.   Figured on the same basis, the retail prices of all consumer goods and services rose 43.4 per cent.

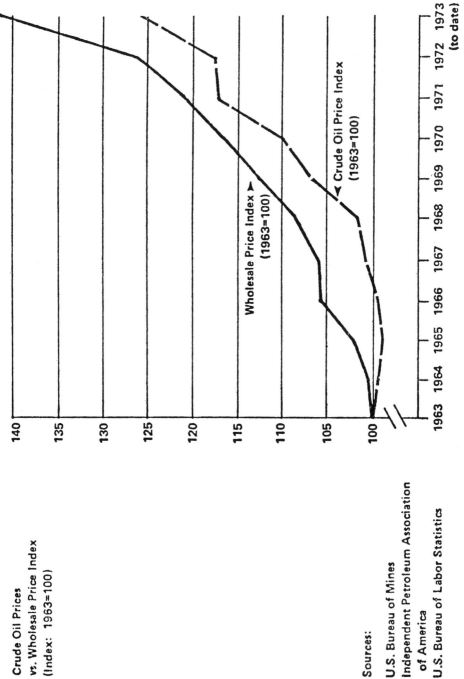

Crude Oil Prices
vs. Wholesale Price Index
(Index: 1963=100)

Sources:

U.S. Bureau of Mines
Independent Petroleum Association
of America
U.S. Bureau of Labor Statistics

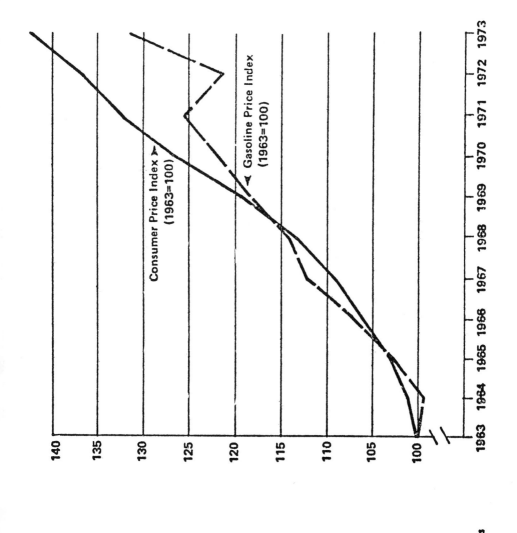

THE INDUSTRY
PRICE RECORD
1963-1973

Regular Grade Gasoline Price
(excluding taxes)
vs. All Consumer Items
(Index: 1963=100)

Sources:

Platt's Oilgram Price Service
U.S. Bureau of Labor Statistics

XII. FREE WORLD PETROLEUM

    1. <u>Foreign oil is neither plentiful nor cheap</u> at the present time:

- World total primary energy consumption increased 47 per cent from 1967 to 1972, and oil continued to provide the lion's share.

- In 1972, oil provided the world with 47 per cent of its energy, up from 40 per cent in 1967.

    2. <u>Total world oil production in 1972</u> was 52.9 million barrels a day, balanced against a consumption of 52.7 million barrels a day. This is not the whole story, however. Even with record production in the Middle East and increased production in Africa, the Free World produced 44.1 million barrels of oil a day -- but it consumed 44.7 million barrels a day, a shortfall of 600,000 barrels a day.

- The three greatest oil consuming areas, the U.S., Western Europe, and Japan produced only 22 per cent of the world's oil in 1972, but they consumed more than 66 per cent.

- Africa and the Middle East, which consume relatively small amounts of oil, produced more than 51 per cent of the world's total in 1972.

    3. <u>Together, Western Europe and Japan took 69 per cent of all the oil produced in the Middle East and Africa in 1972.</u> These areas are and will indeed be competitive for the available oil supplies. In 1972, the combined oil consumption of Western Europe and Japan was over 2.5 million barrels a day more than that of the U.S. In 1967, their combined consumption was 500,000 barrels a day less than that of the U.S.

    4. <u>The ability of the U.S. to meet its needs for oil imports will largely depend on those nations which have banded together as OPEC, Organization of Petroleum Exporting Countries</u>. Four Middle East nations and one Latin American country originally formed OPEC in 1960. By 1972, OPEC had expanded to six Middle Eastern, three African, two Latin American, and one Asian country, controlling more than 80.7 per cent of the Free World oil reserves and 60 per cent of the Free World's oil production in 1972.

- Between 1962 and 1971, the average receipts per barrel by OPEC members increased by more than 60 per cent. Since 1971, OPEC has

negotiated further increases in the price
of its oil, in accordance with previously
established schedules and to compensate
for the devaluation of the U.S. dollar.

XIII. OTHER ENERGY SOURCES

1. Crude oil and natural gas now account for almost 78 per cent of the energy consumed in the U.S. (natural gas provides 32.2 per cent of the total energy, and crude oil 45.5 per cent). A significant cause for the heavy reliance on oil and natural gas has been and is the restrictions -- primarily for environmental considerations -- placed on other energy sources.

2. The natural gas shortfall, like that of gasoline and distillate fuels, is complex in its origin. However, there can be no doubt that a major contributing factor in the decline in natural gas reserves has been unrealistic government policies:

- The low wellhead price for natural gas committed to interstate shipment, set by the Federal Power Commission, has not reflected its true value in relation to other fuels, and has stimulated demand for this clean-burning fuel, while greatly discouraging investment in the search for new reserves; and

- The reduction in the depletion rate for natural gas and crude oil from 27½ per cent to 22 per cent, coupled with other changes in the tax structure in 1969, added more than $500 million annually to the petroleum industry's tax burden -- or roughly the cost of drilling 5,000 exploratory wells.

As consequences of these -- and other -- actions beyond the control of the oil and gas industries, domestic reserves of natural gas have dropped from 20 times yearly production in 1960, to about 11 times yearly production in 1972.

3. Coal, once the major source of America's energy, now supplies only 18 per cent of the energy this nation uses. Yet, it is estimated that coal reserves in the United States could supply our needs for the next 300 years:

- Environmental opposition to the mining of coal, plus federal, state and local restrictions on the use of high-sulfur content coal, have seriously limited use of coal; and

- The coal industry underwent a retrenchment, when nuclear power seemed destined to take over a large share of electricity generation in the 1960's.

It seems unlikely, therefore, that significant additional quantities of coal can become available over the near-term, without substantial changes in the regulatory and labor climate.

4. Hydroelectric power now accounts for about four per cent of our energy:

- While existing facilities in some areas can be enlarged, the expansion would take years to accomplish, and their total contribution would not significantly change the percentage; and

- The limited number of sites close to consumers, coupled with environmental opposition to new dams throughout the country, severely restrict or prevent new dam construction for power generation.

5. Nuclear power, the promise of the 1950's and early 1960's, has failed to materialize:

- In 1968, the Department of the Interior projected an anticipated capacity of 57,000 megawatts of electricity to be generated from nuclear reactors in the U.S. by 1973 -- based on plants under construction, in operation, or programmed at the time of the report;

- In fact, less than one-third of the anticipated 1973 capacity is actually in operation; and

- Nuclear power today provides less than one per cent of our energy needs.

While there is no doubt that nuclear power will be immensely important in the future, it is expected to provide only 10 per cent of the nation's energy in 1985 -- 12 years from now -- because of the long lead-time involved.

6. Other future sources of energy, sources not yet proven economically and technically feasible and for which long lead-times and extensive research and development are required, include:

- Solar power, utilizing the sun's rays to generate both heat and electricity is being explored, with future orbiting satellites envisioned to capture the sun's energy and transmit it to earth stations, where it would be converted into usable energy forms;

- Coal liquefaction and gasification processes, which convert coal to synthetic crude oil and natural gas, are still in the experimental stage, and involve some of the mining problems associated with conventional coal operations;

- Geothermal energy, which uses the natural heat of the earth to produce steam for electric turbines, is now contributing modestly to power generation on the West Coast;

- Shale oil and tar sands are both under development, and while neither has yet proven economically feasible, they do offer significant potential -- again for the future;

- Magnetohydrodynamics, which generates power by forcing electricity-conducting gas through a duct at high speed in the presence of a magnetic field, is still in the first experimental stages, and already evidences problems of disposal of seeding material; and

- Ocean and tidal power, which harness the temperature and height differentials of the ocean to produce power to turn turbines, are basically in the experimental stage -- although one tidal generating facility has been operating in France for a number of years.

7. As promising as some of these more exotic energy sources may be, they cannot be developed in time to meet any significant portion of U.S. energy needs over the short term, when the need for more energy will be the most pressing. Only petroleum offers the answer within the time-frame and environmental requirements of the U.S.